Mark Bloom

TIMES OF TRIUMPH, TIMES OF DOUBT

SCIENCE
AND THE
BATTLE
FOR
PUBLIC
TRUST

Other Titles of Interest from Cold Spring Harbor Laboratory Press

Abraham Lincoln's DNA and Other Adventures in Genetics

Is It in Your Genes?: The Influence of Genes on Common Disorders and Diseases That Affect You and Your Family

The Strongest Boy in the World: How Genetic Information Is Reshaping Our Lives

Also by Elof Axel Carlson

Mendel's Legacy: The Origin of Classical Genetics

The Unfit: A History of a Bad Idea

TIMES OF TRIUMPH, TIMES OF DOUBT

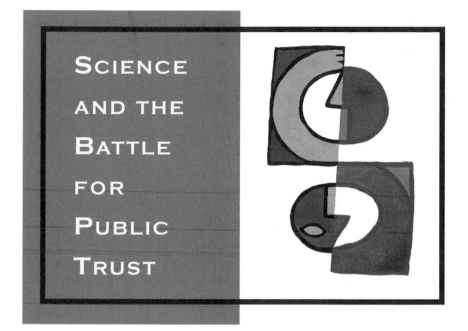

SCIENCE
AND THE
BATTLE
FOR
PUBLIC
TRUST

 ELOF AXEL CARLSON

COLD SPRING HARBOR LABORATORY PRESS
Cold Spring Harbor, New York

Times of Triumph, Times of Doubt: Science and the Battle for Public Trust

Publisher	John Inglis
Acquisition Editor	Judy Cuddihy
Development Director	Jan Argentine
Developmental Editor	Judy Cuddihy
Project Coordinator	Mary Cozza
Production Editor	Patricia Barker
Desktop Editor	Lauren Heller
Production Manager	Denise Weiss
Cover Designer	Paula Goldstein

Front cover artwork: Richard Cook, *Mirrored Faces*, from the Photodisc Green collection. (Courtesy of Getty Images.)

Library of Congress Cataloging-in-Publication Data

Carlson, Elof Axel.
  Times of triumph, times of doubt: science and the battle for public
trust / Elof Axel Carlson.
    p. ; cm.
  Includes bibliographical references and index.
  ISBN 0-87969-805-5 (hardcover : alk. paper)
  1. Science--Moral and ethical aspects. 2. Bioethics. 3. Scientists
--Public opinion. 4. Medicine--Research. 5. Biology--Research.
I. Title.
  [DNLM: 1. Bioethical Issues. 2. Biomedical Research--ethics.
3. Public Opinion. 4. Science--ethics.   WB 60 C284t 2006]
Q175.35.C37 2006
303.48'3--dc22

2006007334

10 9 8 7 6 5 4 3 2 1

Authorization to photocopy items for internal or personal use, or the internal or personal use of specific clients, is granted by Cold Spring Harbor Laboratory Press, provided that the appropriate fee is paid directly to the Copyright Clearance Center (CCC). Write or call CCC at 222 Rosewood Drive, Danvers, MA 01923 (978-750-8400) for information about fees and regulations. Before photocopying items for educational classroom use, contact CCC at the above address. Additional information on CCC can be obtained at CCC Online at http://www.copyright.com.

All Cold Spring Harbor Laboratory Press publications may be ordered directly from Cold Spring Harbor Laboratory Press, 500 Sunnyside Blvd., Woodbury, New York 11797-2924. Phone: 1-800-843-4388 in Continental U.S. and Canada. All other locations: (516) 422-4100. FAX: (516) 422-4097. E-mail: cshpress@cshl.edu. For a complete catalog of all Cold Spring Harbor Laboratory Press publications, visit our World Wide Web site http://www.cshlpress.com.

Dedicated to interdisciplinary approaches to learning:

The Federated Learning Communities
The Honors College
Biology 101-102
All at Stony Brook University

The Danforth Foundation and Lilly Endowment Workshops
on the Liberal Arts
The Lilly Endowment-Poynter Center Workshops on Ethics and Values
The Indiana University Institute for Advanced Study

All of which made this book possible

Contents

Preface

THIS BOOK REFLECTS MANY YEARS OF TEACHING A BIOLOGY COURSE, Biology 101–102, at Stony Brook University, for non-science majors. My motivation was to provide the science that students needed so that they could engage in debate and discussion of controversial issues. I believed it was important, in a democracy, for citizens to be reasonably informed. Part of that course dealt with episodes where harm was done, intentionally or unintentionally, to people, although the participants were usually motivated by good intentions. I also participated in interdisciplinary teaching in the Federated Learning Communities at Stony Brook University which addressed difficult themes (such as social and ethical issues in the life sciences) by bringing together faculty teaching cognate courses needed to explore a common theme. When I served as the founding Master of the Honors College at Stony Brook, I helped design and participate in undergraduate courses that used this comparative approach to learning.

My understanding of ethical and moral systems was assisted by a course taught by David Smith of the Poynter Center at Indiana University which I audited while I was a visiting Fellow of the Institute for Advanced Study at Indiana University. I also benefited from David Smith's 5-year summer workshops on ethics and the liberal arts that he organized for the Lilly Endowment. I believe in the value of interdisciplinary approaches to learning and teaching. The Danforth Foundation provided one such outlet, called the Workshops in the Liberal Arts, which were held in Colorado Springs at Colorado College in the summers. When they moved this enterprise to the Lilly Endowment, I was fortunate to participate for another 15 years as a summer staff teacher. I am particularly grateful to Ralph Lundgren for organizing those workshops and for the talent he used to bring together a faculty committed to exploring the liberal arts. I am also aware of the influence of my mentor, Herman J. Muller, on my outlook, because he did not avoid speaking out on the uses and abuses of science.

Over the years, I have tried to find out why bad things happen when, for the most part, those who worked on the science associated with that harm had only good intentions in mind. It is not easy to probe people's motivations, nor is it easy to reconstruct the influences on people who lived several generations earlier.

The result of that long interest is this book. I hope young scientists, who are unaware of past abuses and errors, will be more reflective about their own values and the values of the people who supervise their work or who pay for their scientific research.

I have chosen examples primarily from the life sciences because that is my area of competence. I am a geneticist and a historian of science. The skills of a scientist differ from the skills of a historian, but both bring an important contribution to understanding how science is used in society.

I thank Dr. A. Peter Gary for his helpful criticisms of the first draft of this book and Owen Debowy for his many discussions of the ethical issues raised in my courses and in these chapters. I am grateful to the Cold Spring Harbor Laboratory Press team for their superb assistance in seeing this book to its publication. I thank John Inglis, Publisher, for his faith in the project; Judy Cuddihy, Acquisitions and Developmental Editor, for her abundant author queries that fleshed out the chapters, not to mention her insights from other works she consulted; Mary Cozza, Project Coordinator, who always had a cheerful disposition encouraging this project; Pat Barker, Production Editor, who converted the corrections and additions into a seamless continuity; Lauren Heller, Desktop Editor; and Jan Argentine, Editorial Development Director.

Introduction

THIS IS A BOOK ABOUT GOOD INTENTIONS and science. Most of the time, good intentions turn out fine and everyone benefits. Sometimes good intentions are accompanied by bad outcomes. Some would call certain bad outcomes evil. In some instances, I would agree with that assessment as a human being, but as a scientist I have tried to find less inflammatory ways of describing bad outcomes. It is difficult to think otherwise than in terms of evil acts when caring people are kidnapped and beheaded, when school children in their innocence are intentionally blown up, when people whose crime is being a Jew, a Gypsy, or a college-educated Pole are killed without having committed any acts against their killers other than being alive. It is an evil act when a government imposes genocide on people of another race, religion, or ethnicity. It is an evil act when humans in large numbers are sterilized without their consent or when they are made subjects of risky medical experimentation that leads to their death.

However, there are far more bad outcomes that are not the work of a few actual or alleged evil scientists who mean to do harm to others in the name of a higher good. There are scientists who had no idea there would be harm from what they did. And there are scores of scientists who have been unjustly accused of committing evil or harmful acts when they actually did good things and their alleged harm is in the imagination of their accusers. I have written this book to look at a spectrum of such cases to delve into the history of the times when these real and alleged bad outcomes occurred and I have tried, where possible, to reveal the motivations of those involved. It is important to do an ethical analysis as well as an historical analysis to seek out the ways people thought, the values they held, and the limitations of knowledge at the time they did their work. This approach permits us to be less judgmental without being forgiving for the harm that was done. One cannot rationalize harm to others as acceptable

1

behavior. It remains wrong for the victims, even if it is rationalized by the perpetrators as done for a higher cause. I hope that this approach will inform readers, especially young scientists, of the way things got out of hand and how they could be more effective in their own careers in preventing future bad outcomes from their work.

In assessing the motivations and behavior of those whose good intentions have been questioned because of bad outcomes, real or imagined, I have adopted the reasoning strategy of being a juror. When one is selected for jury duty, the judge instructs the panel to heed only the evidence and in deliberation to assess the worth of the opposing presentations, to examine the evidence, and to see whether a verdict can be rendered on the innocence or guilt of a defendant. This book is not a set of trials, however, and verdicts of guilt or innocence are not being rendered. Instead, I am presenting the evidence, incomplete as it is in some cases, and trying to render an interpretation rather than a verdict. I am interested in why people do the things they do. I strive to provide the conflicting points of view of the scientists who would mostly consider themselves innocent of wrongdoing and of those critics who believe that they did real harm. The world is not made up of categories of exclusively innocent people and guilty people, or saints and sinners, or scoundrels and victims. The world is composed more often of lots of complex people and sometimes relatively uncomplicated people who are forced to make decisions with insufficient time or experience to think things through.

I am also aware that how people behaved in times past would have been considered acceptable to their peers but condemned by standards of readers today. That is certainly true, historically and culturally, but I believe it is important to understand how harming others was justified then so that we can try to examine our own values and prevent harm to others from our own present and future science.

I have occasionally contrasted science and its values with religion and its values. I do so because most of humanity practices some sort of religion and assigns its values to that upbringing. I neither attack the right to religious belief nor accept its authority on how science should be judged or interpreted.

Those who read this book will include people of religious faith, agnostics, humanists, and atheists. What these groups share is a concern about the values that govern the uses of science in our lives. I have tried to point out where those religious values are pertinent to the outcomes or motivations of scientists and their findings. One cannot ignore the role of religious values in debates over human reproduction, abortion, euthanasia, or stem cell research. To pretend that these values do not exist in society is dishonest and suggests that science works in a moral vacuum. One can disagree with religious views without belittling the faith of scientists or nonscientists who consider religious values part of their lives.

For purposes of disclosure of my own bias, I am a theological atheist (i.e., I live my life without the need for a concept of God) and I am a practicing Unitarian (i.e., I participate in a liberal religion that has no creed but which stresses the importance of living an examined life respectful of human diversity).

PART 1

THE DIVERSE MORAL AND ETHICAL FOUNDATIONS OF BASIC AND APPLIED SCIENCE

SCIENTISTS HAVE DUAL RESPONSIBILITIES. As scientists, they are expected to have ethical standards for conducting research. These include honesty, protecting scientific studies from self-deception, acknowledging sources used for scientific work, and following well-established procedures for gathering data and performing experiments. For scientists who use human subjects, additional ethical standards are required, including the informed consent of the subjects to participate in the scientific studies. Scientists are also citizens, and they are not exempt from the ethical and moral standards of their communities.

Scientists differ, as scientists, in their response to the supernatural. Where the public may have strong beliefs and participation in the supernatural, especially through religious beliefs, scientists have to maintain a strict separation between the material world they study and those supernatural beliefs. It is not always easy for scientists to make that separation consistently. For atheists and agnostics, this is less of a concern than for scientists who consider themselves religious. It is almost certain to be a source of conflict between the interpretation of the universe that science provides and the interpretation of the universe that those who rely on the supernatural provide.

I present in this first section the major philosophical traditions for assessing or guiding behavior. It is important to know these traditions and the historical basis for those views. Science does not work without some sort of moral and ethical guidance. Scientists, like all other professionals, have to be aware of the consequences of their work. When the public trust in science is betrayed, all science suffers, and excessive restrictions may be imposed on science through regulation, policies that inhibit research, legislation that defines what science can or cannot do, or the amount of financial support provided to carry out the work that scientists hope to do.

As shown in the historical examples used in this analysis of real and alleged bad outcomes of science, the failures involve both personal mistakes or moral lapses by scientists and a policy of publicly sanctioned behavior (especially during wartime or during the threat of war) that can betray both science and the public. It is far easier for scientists to accept and evaluate such assessments about the use and abuse of science than it is for scientists to respond to the public when public belief favors a worldview inconsistent with science. Scientists have limited options for their response to conflicts of worldviews. They can ignore those supernatural beliefs; they can confront those beliefs; or they can make nonhostile efforts to educate the public. Scientists are more likely to respond to such public attacks on science when students are taught supernatural explanations in science courses that are imposed by local or state law or by institutional policy. Most scientists would vigorously resist being asked to teach supernatural explanations for work in their own fields.

1

Why Science Is Sometimes Perceived as Evil

I AM A SCIENTIST AND IDENTIFY MYSELF AS A GENETICIST. I have had a wonderful career teaching the life sciences to undergraduates, graduates, and medical students. My love for science led me to study its history, and I have a deep appreciation for the history of my field and have written several books about it. I know that science has contributed to the good of humanity through the extended longevity of humanity in the 20th century; through the control we have over our lives that has replaced fatalism; through its satisfying explanation of much of the universe; and through its many tools that enable us to construct civilizations.

Many students, who were willing to tell me so, shunned science and feared science. They looked upon it as alienating, threatening to their religious beliefs, and capable of monstrous evil. They saw science as cold and aloof from the life in the humanities they preferred. Yet few of these students would have wanted their world to return to an era before, for example, germ theory, in which they would have been vulnerable to a premature death. They would not want taken away the technology of printing, telephones, television, jet transportation, automobiles, computers, all-year-round fresh fruits and vegetables, and public health. They are ambivalent about science more than opposed to it. They feel that science has let them down through its bad outcomes.

The Bad Outcomes of Science: A Litany

What are these bad outcomes? The first issue is what I would call the Frankenstein image of science. In our popular culture, we often refer to

Mary Shelley's novel of 1818, *Frankenstein, or the Modern Prometheus*, in which Dr. Frankenstein creates a soulless being who feels unable to relate to humanity and who lacks the restraints we usually muster when we confront disappointment. Frankenstein is perceived today as the scientist motivated by a shallow value (to see whether he can synthesize human life by using electricity to reanimate dead flesh) and whose creative act leads to bad outcomes. The Frankenstein image has degenerated into a mad scientist image in movies and in the public imagination. In fact, the Frankenstein novel is a replay, as the usually omitted subtitle suggests, of the stories of Prometheus, Pandora, and Eve, who all chose the acquisition and use of new knowledge to change humanity. Humanity has always feared the unknown and the future because life was largely unpredictable and humans had little control over their own lives.

In the absence of science, the insecure found some solace in the promise of religion. The virtuous in this worldview are rewarded, if not in this life, then in an afterlife. Without science, one's choices are usually fatalism, stoic resignation, or prayer. But the flip side of providing new knowledge in these legends is punishment. For Adam and Eve, it is expulsion from Paradise, and a life of pain and hard work. For Pandora, it is a release of the ills of humanity. For Prometheus, it is his own liver being devoured every day of his life. Because of this Frankenstein image, the scientist coexists in popular culture as hero, villain, and martyr.

A second issue for the bad outcomes of science involves its crimes against humanity. We think of the Holocaust and the efforts of Nazi scientists to design gas chambers, to find an effective gas to kill inmates with a minimum harm to the killers, to do life-threatening or murderous medical experiments on human subjects, to classify who should be killed immediately and who should be sent into slave labor, and to justify the perceived inferiority and vermin status of Jews and other Nazi-alleged inferior races and ethnic groups. We think of nuclear war and the hundred thousand victims each in Hiroshima and in Nagasaki who were sacrificed to hasten the end of a war at its terminal stages. We think of the eugenics movement and its depiction of social failures as biologically degenerate, meriting compulsory sterilization. We also think of scientific racism with its long history of classifying humanity in hierarchies and assigning negative attributes to those lower down in the scale of accomplishments. In all these cases, humans were deprived of their humanity and forced into slavery, deprived of having children, or even killed to justify a view that their lives were expendable or worth less than those of their rulers.

A third issue involves the bad outcomes of unregulated or ineffectively regulated science. We reacted to tainted foods that caused food poisoning during the Spanish-American War and enacted laws for the inspection of our foods. We demanded engineering inspections and oversights for new skyscrapers, railroads, ships, dams, bridges, and other facilities that are constructed and used near populated areas or that carry large numbers of people. Each failure has led to new regulations. In some instances, failures have a way of setting back or ending an applied use of science. We think of nuclear reactor failures and the setbacks to the nuclear reactor industry. We think of the NASA Columbia and Challenger space shuttle disasters and the long delays that ensued in the manned space program. We demand that answers be found when airplanes crash or ships sink. If we have rejected fatalism in nature, we also seek to banish future accidents associated with known defective design, construction, or operation. We properly deplore the occurrence of, and seek to prevent, failures that led to the deaths of thousands of people in 1984 who happened to live in Bhopal, India, near a poorly designed chemical facility that failed to protect the public.[1] We will not allow a relaxation of regulatory standards that were imposed on the drug industry in many countries after the malformation of some 8,000 babies whose mothers used thalidomide for morning sickness.

The fourth issue science has to face is its conflict with worldviews that are false or inconsistent with science. We think of Galileo's demonstration of the solar system using a telescope to identify Jupiter's moons, the moon's craters, Venus's phases, the existence of sunspots, the rotation of the sun, and the presence of rings around Saturn. His fellow astronomers, as well as his church, sought to silence him for threatening a naïve view of the universe that saw the earth as the largest object in the universe and at its center, with everything else moving in concentric rings (with accompanying epicycles) around it. Galileo was forced to deny his own arguments and lived in house arrest for the last years of his life. Charles Darwin was fortunate that his views on evolution by natural selection were proposed in the mid 19th century. He might have suffered a fate worse than Galileo had he lived in the 17th century and made that proposal. Even if there were no punishing consequence to his publication of *The Origin of Species*, Darwin was reviled by many of his readers for introducing a materialist view of life in which God played no essential role. Darwin was aware of this possibility and looked with anguish upon his own theory as murderous to the religious views of creation then in vogue. These works of Galileo and Darwin are not applied sciences. They are basic sciences where facts, observations, and testable theories are at odds with popular (usually religious) beliefs about the universe. Whether scientists like it or not, much of their work is seen as immoral or hostile to the religious views of others.

The Tasks of Science in Response to Its Critics

It would be difficult to deny that science has lent its support to developing weapons of mass destruction, to supporting the evil goals of dictators and misguided governments, to rationalizing bigotry, to offering badly designed or carelessly used facilities that failed, and to backing iconoclastic ideas that have shaken people's faiths. Science is a human activity capable of making errors and subject to abuse. It also provides a worldview at odds with superstition, the supernatural, and false descriptions of reality. I hope to assess those charges against science, and I believe I can sort out the fair claims from those that are spurious. If there are legitimate grievances, then society or science itself has to address a way to prevent harmful outcomes from the applications of good intentions. This will require a combination of a historical analysis of what happened and an ethical analysis of the motivations and intentions as well as the behavior of the scientists involved in these episodes. I have chosen a spectrum of failures as well as a selection of controversial issues in contemporary science where the prospects of harmful outcomes are unjust or unproven. My hope is that a review of these events and cases will lead the young scientists reading this book to become more reflective about their own motivations and values. For most scientists, science has led to good outcomes that have reinforced both the pleasure of doing science and the satisfaction of contributing to human benefit. That is the ideal of science that we cherish.

Notes and References

1. Charlotte Crabb, "Revisiting the Bhopal Tragedy." *Science* **306:** (2004): 1670–1671. Crabb points out that in 1987, about 44% of 865 pregnancies ended in miscarriages. The gas cloud was probably laden with hydrogen cyanide. At least 3000 people died within days after their exposure. The legal case in 2006 is still being fought in Indian and U.S. courts, almost 20 years after the accident occurred.

2

The Tools of Judgment
Ethics and Moral Traditions

As long as humans have lived in social groups, they have established rules of normal behavior. Cultural anthropologists have found few universal rules of behavior, but all societies regulate killing, theft, betrayal, harm to others, social interaction, and marriage. Written laws date back three or four millennia. When bad outcomes occur, those who are victims seek justice. They look for punishment of wrongdoers; they look for prevention of the same thing happening to others; or they seek compensation for the harm inflicted on them. Most citizens today would agree that courts are the most neutral and formal settings for making such judgments about bad outcomes. But courts cannot prevent or foresee bad outcomes from new science. That has to be assessed through regulation by scientists themselves; through regulation by legislative acts; or through intense scrutiny by the press in response to real or imagined concerns of those who fear new science.

Rights-based and Outcome-based Ethics

When we make decisions based on our values, how should those values be classified? The two major categories of ethical decision-making are rights-based and outcome-based.[1] Rights-based ethics (also called deontological ethics by philosophers) are usually assigned to religious traditions such as the Golden Rule (Do unto others as you would have others do unto you) or the Ten Commandments (God-ordained rules of behavior) in the Judeo-Christian-Moslem faiths. Rights-based ethics can also be based on reason without invoking the supernatural, as Kant pointed out in his self-testing of ethical decisions. In the Kantian scheme, he asks how a rational person would respond if asked, "Is it permissible for me to take your money or property without your consent?" The rational person would say no. If *all* rational persons reply the same way, then it is reasonable that theft is universally rejected. The same question can be applied to more specific issues: "Is it permissible for me to do medical experiments on you without your consent?" "Is it permissible for me to sterilize you without your consent?" Note that the Golden Rule and Kantian ethics are similar if not identical. Also note that a scientist who is an atheist can take a moral position because reason, not religion, determines Kantian ethics.

We also practice a very different system of ethics based on outcomes. In the late 18th and 19th centuries, this way of making decisions was called utilitarianism. In utilitarian ethics, one does "the greatest good for the greatest number."[2] This is a democratic ethics. It assumes a fairness and a majority rule for assessing benefits. It presupposes some sort of mental arithmetic in which we put a higher or lower value on the good intention or the bad outcome. We invoke the phrase "outweighs" in assigning that judgment, as in "the dropping of the atomic bombs in Hiroshima and Nagasaki to end the war quickly outweighed the cost in lives to innocent Japanese men, women, and children." In our mental arithmetic, we imagine 300,000 deaths for allies invading the Japanese mainland in a campaign that might take 6 months. Against this, we imagine some 20,000–100,000 lives lost in the atomic bombings, and we use this weighting of outcomes to justify the "saving" of those Allied lives (not to mention the many hundreds of thousands of civilian and military Japanese lives that would have been lost in a prolonged invasion).

For those who favor rights-based ethics, the killing of innocent people is absolutely wrong. Alternatives must be sought based on those universal rights. In reality, we use both ethical systems, and we fudge our consciences with rationalizations and exceptions. One of the Ten Commandments says "Thou shalt not kill." But in our own minds, it's OK to kill in self-defense; it's OK to kill a combatant in war; it's OK during war to kill innocent people who happen to be in the wrong place (we call it collateral damage); and it's OK to kill those the state deprives of life for crimes (heresy, counterfeiting, kidnapping, picking pockets, murder, rape, bestiality, adultery, homosexuality, atheism, blasphemy, or disobedience—all depending on the country and the times). Ethics is not limited to deontological and utilitarian interpretations. Some systems of ethics are based on virtue (Aristotle), contractual consensus, or goal-directed activity. The two most frequently invoked in the scientific controversies in this book are the rights-based and outcome-based kinds. Because teleological thinking (goals as final causes) and intrinsic virtues are not part of most scientists' habits of thinking, these are rarely invoked in the debates about intentions and outcomes in science.

We are often unaware of the inconsistency of our own ethical beliefs and create what I call "constellations of values."[3] A moral conservative in the U.S. at the start of the 21st century in general favors the death penalty for criminals, the regrettable necessity of killing civilians when engaged in war, and the rejection of any form of reproductive decision-making by a woman about those contraceptive practices that involve abortion or artificial means to prevent implantation of a fertilized egg. Such conservatives believe that they have absolute rights-based values and see no contradiction to the commandment "Thou shalt not kill." A moral liberal in general favors the right of women to make reproductive decisions, including abortion and birth control; the need to avoid killing civilians in times of war; and the elimination of the death penalty as cruel and unusual punishment (what could be more cruel than taking away another person's life?). Those moral liberals do not equate abortion and preimplantation devices that destroy fertilized eggs as killing. Each side has its constellation with inconsistent values.

The reason for this inconsistency is that life is complex and there are different standards we use, mixing both utilitarian and rights-based values. Often the constellations of values we adopt are derived from, or include, religious traditions. Sometimes they become political beliefs as well. Some people want their particular set or constellation of values to be universal, and they look at departures in other groups as immoral or criminal. New technologies also force older value systems to adapt to the new. To make those constellations fit, a lot of supplementary reasoning is used. Distinctions are made between the natural and the artificial; distinctions are made according to motivation. A theological or philosophical concept called natural law may be invoked. For the moral conservative, it is artificial or unnatural or violates natural law (and hence is not permissible) to use donor sperm when the husband is infertile. But it is permissible to use a donated kidney when the recipient is suffering from kidney failure. Kidney failure is considered by conservatives and liberals alike to be a disease. Sterility, for the moral conservative, is not considered a disease, and thus, using someone else's sperm when the male is sterile is not a morally permissible treatment. For someone not beholden to natural law theory, the difference is one of allowing a donor's somatic tissue to be used in a treatment but not a donor's germinal tissue. That strikes the needy person as arbitrary. Thus, it makes no sense medically or scientifically, and it only makes sense from a particular religious or philosophical perspective.

Evaluating Ethical Systems

It is difficult for any ethical system to apply universally to all situations. We recognize myriad exceptions to our generalized ethical positions. We have to, or life would be very hard to live. Virtually no one wants an absolute ban on killing humans despite the revulsion most people feel to any death that is inflicted by one human on another. We distinguish intent when killing occurs. We convict some of murder and some of lesser charges such as manslaughter or negligent homicide. Some killers are not convicted because jurors see no crime committed and believe an accident occurred or that the homicide was justified. In these exceptions to our general rule, we are often using utilitarian values to justify our decision.

It is not just for classifying killing that utilitarian ethics become murky. The greatest good for the greatest number can result in a tyranny of the majority. If in 1790 there were a lot of poor people and augmenting the poor laws (public charity in Great Britain) outweighed the burden of increased taxes to the wealthy minority, that would be a utilitarian application offensive to a conservative. However, if in the 1990s the middle class in the U.S. greatly outnumbered the poor, then cutting off or dramatically reducing aid to the poor, thereby easing the tax burden of the middle class, should provide the greatest good to the greatest number. That application of a utilitarian interpretation would likely please a conservative.

A second feature of our ethical beliefs is that they are largely learned by mimicry as we grow up. We see the behavior of those who raise us and those who are our neighbors and family acquaintances and relatives. If there is substantial honesty and integrity in the way they treat each other, children pick this up without being instructed by school or church. We say that we learn our ethics at our mothers' knees.[4] Some ethics teachers are pessimistic that what they teach makes a difference in their students' lives, especially at the college level. By that time, patterns of behavior are difficult to change, and what one learns for a school examination does not necessarily alter one's basic behavior.

Professions have their own rules of ethics. For physicians, there is the universal Hippocratic oath "to do no harm." Some physicians see their role as limited to that of a healer—they diagnose and treat illnesses. Some see their role as broader and include aspects of teaching, counseling, and advocating preventive medical habits. For a long time, medicine was paternalistic, and physicians withheld bad news and gave as little infor-mation as possible to patients, assuming their patients would get depressed or give up the fight to live with a disease if they were told the truth.[5] After the 1970s, things changed. Patients are treated as having their own autonomy that must be respected. Now patients have to struggle with the knowledge of chronic disease, terminal cancer, heart failure, or the limited prospects of organ failure.

For most scientists, there is no oath to take. They do not usually deal with people as subjects. Their attention is to an external reality they hope to explore, describe, test, and manipulate. Their ethics is primarily one of integrity, respect for truth, designing controlled experiments to prevent personal bias (conscious or unconscious) from warping their work. Unlike physicians, scientists rarely have to consider the impact of their work on those who are not scientists. This creates a problem of over-looked responsibility. The vast amount of science is not university research detached from application. It is science put to application. The defense industry uses thousands of scientists to solve complex problems in making weapons. The space industry depends on scientists to design rockets, space stations, satellites for communications, and advanced tools for exploring the earth, planets, stars, and galaxies. The pharmaceutical industry relies on scientists to design and find new drugs. The agricultur-al industry depends on scientists to use genetics, plant physiology, cell biology, and many other tools to design useful crops resistant to disease or capable of growing in inhospitable places.

Who should take an interest in the possible bad outcomes of all this research? Medicine is required by numerous regulations and laws to pro-tect the public it treats. Scientists sometimes argue that their job is to solve the technical or scientific problems, and it is not their responsibility how that knowledge is put to use. But how valid is that claim? In Nazi Ger-many, if scientists were asked to design crematoria for the burning of numerous bodies using the most efficient way to reduce fuel costs, would not this strike some scientists as a morally troubling assignment? If scien-tists were asked to design a nerve gas that would kill rapidly and penetrate all but the most sophisticated anti-gas outfits, would that not disturb some scientists? Just because the scientist is not the person who delivers those tools of death to others does not mean that moral reflection in the scientist's life is limited to mundane issues.

In our discussion of ethics and values, we have assumed that the source of our standards comes from religion, custom, or reason. In the

past 40 years, there has been a flourishing school of evolutionary psychologists who have used the terms sociobiology, comparative ethology, or
biological determinism to describe their work. The premise of their
claims is that much of human behavior is not learned but reflects genetically programmed ways we respond because of our evolutionary history.[6]
There is much debate about these claims. Much of the evidence is indirect
and comes from animal studies, especially among the primates. Earlier
attempts to justify human behavior through an imagined evolutionary
past included both pessimistic views of human behavior (selfishness,
greed, bigotry, territoriality, exploitation) and optimistic ones (the evolution of cooperation among humans, or mutual aid, as Prince Peter
Kropotkin called it in the 19th century).[7] The danger of these studies is
that they may be used to construct allegedly objective characteristics of
human nature which are not based on religious traditions. They could be
used to justify or tolerate a variety of legal but unpleasant behaviors in
humanity, including male dominance, male promiscuity, territoriality,
desires for power, aggression, and even racism. It is still premature to call
such traits, identified through evolutionary psychology, as partially or
totally genetically driven until a genetic analysis shows that this is so or
until the genes favoring them are mapped and sequenced. Many evolutionary psychologists tone down the implications of their work by
endorsing an environmental influence on a range of possible behaviors.
The genetic component is seen as probabilistic and not rigidly determin-
istic. This makes it difficult to assess what the genetic component actually does. It may turn out that the evolutionary psychologists are correct; it
is also possible that there will be few, if any, such genetically programmed
human behaviors. Whatever the truth of such biological determinism may
be, our laws will still hold people accountable for those acts society con-
siders criminal or harmful to others.

Humans judge one another by the outcomes of their behavior. Good
intentions do not relieve those who cause harm from both legal and public condemnation. Scientists may be uncomfortable when others use terms
like evil, immoral, or depraved to describe the harm done by the applications of science. Some scientists walk away from personal responsibility,
believing that they only provided the know-how and were not authorized
to make the decisions that harmed others. That is generally true for those
whose research is not intentionally applied and who were only seeking to
understand nature. It is almost certainly false for those who work to apply
science for human use.

Notes and References

1. The Golden Rule may be expressed in two major ways: "Do unto others as you would have others do unto you" and "Don't do unto others what you would not want others to do unto you." The first puts a lot of obligations and an uncertain inclusiveness of doing this (e.g., should it include giving to charity, loving one's neighbor, or other good intentions?). The second puts limits on behavior and thus keeps the focus on preventing bad outcomes. There is a third way to express this relation, but it may not qualify as a Golden Rule: "Do unto others as they do unto you." This is tit-for-tat or eye-for-an-eye ethics, which does not take into account motivation (What if it was an accident? Do I really want the eye ripped out of a person who was careless or unlucky?) We usually consider the eye-for-an-eye value outmoded and vengeful. The Golden Rule has many variants in its Greek (5th century B.C.E.), Confucian, Jewish, and Christian origins.

2. John Stuart Mill and Jeremy Bentham are the two most cited advocates of utilitarian ethics. Rights-based ethics permeated the thinking of Charles Dickens, who parodies utilitarian ethics in his novel *Hard Times* (1854; reprinted, 1990, W.W. Norton, New York). Utilitarian ethics is often classified today as consequentialism.

3. There is nothing except the accident of location that produces an astronomical constellation. We take bright objects and link them into a unit. The stars in a constellation could be of different ages and different distances and different sizes. The names and shapes of the images also differ from one culture to another. If we were to move the earth a few light years from our present location, many of our constellations would disappear and new ones of different shapes would be seen. I use this term to imply that it is some consensus of theology and politics that creates a constellation of conservative or liberal values.

4. This was forcefully expressed by Gilbert Nielander at the Lilly-Poynter Center workshops on ethics and the liberal arts. Most faculty participants disliked his view because they felt their teaching was important and changed students' views. I felt that way, too, but when I reflected on the Watergate trial of Jeb Stuart Magruder, I had to rethink my view. Magruder's ethics teacher, William Sloane Coffin, described him as his most brilliant student.

5. I recall when my eldest brother was dying of lung cancer in 1972 that his physician demanded that neither his wife nor his siblings should mention to him that he had lung cancer. "To do so," he claimed, "would deprive him of hope and the will to live." That is a paternalistic judgment, as evidenced by J.B.S. Haldane's reaction when he was deceived by his physician and told that his bowel cancer was cured after surgery. When he realized he was terminally ill, he berated his physician for depriving him of an opportunity to set his priorities for research and writing.

6. Edward O. Wilson, *On Human Nature* (Harvard University Press, Cambridge, 1978). Steven Pinker, *The Blank Slate; The Modern Denial of Human Nature* (Viking, New York, 2002).

7. Peter Kropotkin, *Mutual Aid: A Factor of Evolution* (1902, reprinted 1987, Freedom Press, California). See also Roderick Gorney, *The Human Agenda* (Simon and Schuster, New York, 1972).

PART 2

THE ROLE OF THE STATE IN SHAPING SCIENCE

S CIENCE HAS ALWAYS DEPENDED ON SUPPORT to do its work. Private philanthropy, royal patronage, self-subsidy by the independently wealthy, industrial wealth to provide research opportunities for those industries, and state support are part of the economics of science. State support is more recent in the history of science. For scientists, the ideal relation is a state agency whose members are civilian scientists who use peer review to support research. The U.S. National Science Foundation and the U.S. Public Health Service's National Institutes of Health are among the most admired agencies by scientists because they are relatively free of restrictions by the government on what can be studied and how scientists should conduct their research.

When governments intervene in science and dictate what types of science should serve the state, what types of science should be purged from state support or banned altogether, the results can be traumatic. Although the Lysenko affair in the former Soviet Union lasted from 1935 to 1960 and had destructive effects on the teaching and study of classical genetics, I have chosen, for this section, the development and practice of life sciences in Germany from 1932 to 1945 to illustrate the malignant influence that is possible from the state's endorsement of science that supports its ideology.[1]

Compared to the direct influence of the state in the Soviet Union in the Lysenko controversy and the direct influence of the state in support of race hygiene ideology in Nazi Germany, the influence of the state on science in the U.S. was far less severe. In the U.S., there were state laws that permitted the sterilization of individuals described as "unfit" or degenerate and whose reproductive cells were considered genetically defective. I examine four individuals who were involved in the eugenics movement: Charles Benedict Davenport, the acknowledged leader and theoretician of the eugenics movement in the U.S.; Harry Laughlin, Davenport's "point man" who led the effort to obtain legislation favorable to eugenic sterilization and restrictive immigration; Sir Francis Galton, whose eugenic influence on the U.S. was indirect and more benign than harmful in intent; and Harry Clay Sharp, an idealistic physician eager to apply hygiene to benefit the health of his own patients and to extend the alleged benefits of that hygiene to society by advocating his fellow physicians to act as lobbyists to enact state sterilization laws.

I chose those four because there is sufficient documentation to reveal their personalities, their beliefs, and the social circumstances that justified their good intentions. It would be difficult to deny that sterilizing people without their consent was a bad outcome, at least for those 40,000 who were forcibly sterilized.

Some readers may reject my assessment of eugenicists Charles Davenport and Harry Laughlin as having personalities like war criminal Adolph Eichmann (described by Hannah Arendt's memorable phrase, "the banality of evil"). They may feel they are not very different from Francis Galton and Harry Sharp. Others may feel all four do not compare at all to Eichmann. That is certainly the reader's privilege. My assessment is based on the deep insecurities of Davenport and Laughlin, which they share with Eichmann. Galton and Sharp, I believe, were motivated by an idealism, and their failings (an amalgam of unexamined bias and neglect) are more difficult to recognize or condemn. Galton spent most of his life promoting eugenics for geniuses like himself. In his books, he was not preaching the elimination of the unfit but the propagation of those with talents and genius. Sharp was filled with idealism as a young physician who saw public health as his mission. Neither Galton nor Sharp sought to do harm to others as a path to personal advancement. I believe Davenport and Laughlin did.

Note and Reference

1. For a thorough account of the Lysenko controversy, see Nils Roll-Hansen, *The Lysenko Effect: The Politics of Science* (Humanity Books, Amherst, New York, 2005). Roll-Hansen provides a superb analysis of the scientific dispute and the gradual shifting away from a conflict among scientists to a state-supported view of science based on politics and ideology.

3

Evil at Its Worst

Nazi Medicine and Biology

N O EVIL ACT IS SO OUTRAGEOUS THAT IT DOES NOT find some sort of support. Humans have enormous capacities for looking the other way, rationalizing, or denying that what is evil should not be judged as such. There are many who prefer to see any moral situation in "shades of gray," almost as a continuum without boundaries. That gray spectrum is probably true of most societies (which are mixtures of many behaviors) and probably true of most self-assessments, even by those convicted of the most heinous crimes. Today, the Holocaust and Nazi human biology under the Third Reich (1932–1945) is usually recognized as evil science (or at the farthest end of the gray spectrum) by virtually all of humanity except Holocaust deniers, anti-Semites, and Nazi sympathizers. At the time of its formation, the race-hygiene movement in the Weimar Republic and pre-World War I Germany was not as virulent as it became under Nazi Party leadership. Why is this so?

Race hygiene was conceived by Alfred Ploetz (1860–1940). Ploetz was inspired by the social Darwinist movement developed by Herbert Spencer and by the eugenics movement developed by Francis Galton. He was also sympathetic to the views of Ernst Haeckel, Germany's leading popularizer of evolution and social Darwinism. He combined ideas of public health and the hygiene movement, championed in Germany by Rudolph Virchow, with the idea of a German people (Volk) that was part of the Romantic Movement in Germany. The Wagnerian operas fostered that movement by celebrating a mythic German brotherhood having hereditary ties to antiquity. For Ploetz, it was not the individual for whom eugenics was to be applied, but the Volk. The Volk was seen as a breed or biologically based culture. It was not humanity (Galton's ideal) but the Volk or race that needed protection through the purging of its unfit or less desirable members and by encouraging the Volk to purify itself through selective breeding. Ploetz was not an anti-Semite. He was a German nationalist, and he felt each race owed to itself the necessity of adopting race hygiene.[1]

Adolph Hitler shaped National Socialism or the Nazi Party. It was an amalgam of Hitler's distaste for democracies, a virulent anti-Semitism, a desire to restore Germany to greatness through a revived military, a socialist control of health services for the German people, a celebration of the German Volk as superior to all others, a desire for revenge from the humiliation of defeat in World War I, and a belief that authority should be top down—the general public lacking both knowledge and the will to act. None of these components was original, but Hitler had great political skill in attracting like-minded supporters who wanted an alternative to the paralysis, defeat, impoverishment, and confusion that plagued Germany after the Versailles Treaty.[2]

Hitler and the Nazi Party created a state based on a biological model. Among its components was the belief that the state was like a human body and subject to cancers and other diseases. Just as a physician is needed to treat the illness of the individual body, a state in this model is obligated to cleanse itself of diseased components. The model goes back to antiquity, and in its modern form was detailed by Herbert Spencer in his *Social Statics* (1855), the founding document of what later became Social Darwinism.[3] A second component was the belief that almost all cultural differences in humanity were biologically based. Each nationality had its own alleged hereditary basis for its characteristics. These ideas were derived from the work of Joseph Arthur Gobineau who founded scientific racism, William Parr who coined the term anti-Semite and founded the modern anti-Semitic movement in 19th-century Europe, and Houston Stewart Chamberlain (Wagner's son-in-law) who extended Gobineau's ideas of the superiority of the Aryan race and who was an anti-Semite. The final biological component was the eugenics movement, endorsed and extended by the Nazi Party, which used American-based restrictive marriage and compulsory sterilization laws as models for its own infamous Nuremberg racial laws after Hitler came to power. It is important to note that most of these components predated Hitler's formation of the Nazi Party or even his own birth.

There is considerable disagreement on when all these ideas came together and how they were modified during the 13-year Reich into a juggernaut of evil. That evil I identify with the use of the state to sterilize people against their own will; the use of the state to order secret euthanasia on Aryans who were mentally retarded, psychotic, or victims of birth defects; and the use of the state to deprive Jews (and a much smaller cohort of Gypsies and others deemed undesirable) of their rights as citizens (including occupation, housing, education, marriage, and eventually life itself). The obsession with a Final Solution, culminating in the Wannsee Conference on January 20, 1942, was the terminal stage of an ascending horror that the Nazi Party inflicted on the families of those whose relatives were deemed unfit to live and those who were stripped of their humanity for the alleged biological pathology of being Jews or Gypsies.

To achieve this purging of the German Volk (later extended to German-occupied Europe), the Nazi Party relied on experts who could be recruited to sterilize and kill, to serve in courts to legalize the sterilizations, and to serve in hospitals to decide who was to be euthanized. The major scholars involved in this state-dictated biology included German geneticists, anthropologists, psychiatrists, and physicians. The historical record is clear that virtually none of these accessories to state ideology was forced to do the work they did. They were in some instances architects of the ideology or designers of the machinery for implementing that ideology. In other instances, they were motivated by "good intentions"— the cleansing of the German Volk, an ideology they found intellectually honest and emotionally satisfying. For the most part, they did not consider what they were doing as evil.

Also involved in these acts of evil were non-scientists, including the top leadership of the Nazi Party (Hitler, Goering, Goebbels, Himmler); specialists in the military, the state police (SS and Gestapo), and top party staff (Heydrich, Eichmann, and the designers and commandants of concentration camps). Those individuals usually lacked medical or biological training, but they were convinced of the scientific legitimacy of the biological models their academic colleagues provided.

Motivations of the Theoreticians

Two books have investigated the motivations of these German scientists and physicians. Robert Jay Lifton's *The Nazi Doctors* (1986)[4] and Benno Müller-Hill's *Murderous Science* (1988)[5] both present interviews with surviving participants or those who knew them. Müller-Hill's subjects were geneticists, anthropologists, and psychiatrists; almost all of them were professors in good standing at the time of the formation of the Third Reich. They provided the theories used to shape the biological, racial, and medical ideology of National Socialism. In all likelihood, none of them personally killed patients or ordered the killing of those they deemed unfit. All of them after 1933 approved of policies based on their theories including the forced segregation, banishment, and forfeiture of citizen's rights of the Jews. They did not consider themselves anti-Semites (at least before 1933 or after 1945) or sympathetic to anti-Semitism. They trained physicians and other professionals to identify Jewishness among alleged half-Jews, Gypsy traits (almost all Gypsies were mixed with other ethnic groups), psychosis, mental retardation, and a variety of disordered personalities. They claimed they complied with laws of the state and that their role was that of scientists providing impartial evidence to be used by the state according to its laws. None acknowledged that they knew of the Holocaust or that killings were done for reasons other than the strictest diagnoses to provide euthanasia for hopelessly ill patients.

Quite different are the responses to Lifton by the physicians he interviewed. These were the foot soldiers of the Holocaust and euthanasia movement. They knew what was happening because most of them worked at the camps or the hospitals where the killings took place. They did the selections of who would be gassed and who would be worked to death. Some physicians used these patients for experiments that they felt would be helpful to medical science and human biology. Although they knew this was medically immoral and illegal before the Third Reich, they accepted their work as medically necessary and legal under the Third Reich.

Both Lifton and Müller-Hill try to assess those behaviors and motivations. Müller-Hill emphasizes the role of denial, evasion, silence, and public policy as shields the theoreticians used after the war when they had to justify themselves to their colleagues, country, war tribunals, and themselves. Lifton emphasizes the role of "numbing" (getting used to unpleasant work through repetition), displacement (he coins a concept of "doubling" where an alter ego personality coexists with one's normal self), and released sadism (many of those drawn to the work or who stayed with it had disordered personalities). In both categories of major perpetrators, some of the worst offenders committed suicide or managed to escape, but a considerable number were acquitted or never indicted, and most of the professors who created the justifications for mass murder ended up working at their old university positions after the war.

Rationalizations to Offset Evil Behavior

Müller-Hill's indictment of the theoreticians is chilling: "None of them had incriminated themselves. Most of them had declared that they had not known anything of the mass murders of mental patients, Jews, and Gypsies; that they had not known anything of the events that had linked their academic institutions with the murderers; that there were no traces of anti-Semitism in their field of work; that either they had not joined the Nazi Party at all, or that they had been forced to join; that their best friends were Jewish or half-Jewish; that, as pure scientists, they had never handled expert reports on racial origins; and that they had given little or nothing to the general context of these events."[6] Müller-Hill identifies seven stages in a progression from 19th-century ideology to the genocidal ideology of the 1940s:

- A biological basis for human diversity (especially what Spencer developed and which came to be known as Social Darwinism)

- Classification of the socially unfit (mostly paupers, psychotics, the mentally retarded, and recidivist criminals)

- Classification of racial inferiors (with Nordic superiority)

- Sterilization of the unfit (beginning in 1909 in Indiana)

- Isolation of the unfit (through asylums, ghettoes, or deportation)

- Euthanasia of the unfit (alleged life unworthy of life)

- Mass killing of the unfit (the Holocaust or Final Solution)[7]

The architects of Nazi biological ideology include Alfred Ploetz (1860–1940), Eugen Fischer (1874–1967), Fritz Lenz (1897–1976), Otmar von Verschuer (1896–1969), and Ernst Rüdin (1871–1952). Ploetz coined the term, race hygiene, in 1905 and was founder of the racial hygiene movement in Germany. Fischer was a geneticist especially interested in human genetics (but considered himself an anatomist by training). Lenz also pioneered studies in human genetics. von Verschuer was an anthropologist interested in the genetics of racial differences. Rüdin was a psychiatrist. All were professors, and most were Directors of the Kaiser Wilhelm Institutes (later Max Planck Institutes) noted for distinction in scientific achievement.

In 1925, Adolph Hitler read the second edition of a text in human genetics (*Grundriss der menschlichen erblichkeitslehren und rassenhygiene*,

J.F. Lehmann, Munich) written by Erwin Baur, Eugen Fischer, and Fritz Lenz. It can be considered the first medical text in human genetics. Baur (1875–1933) was a plant biologist whose work on cytoplasmic inheritance in snapdragons was internationally respected. He was largely apolitical and did no direct work in human genetics and thus played no known role in forming Nazi ideology. He received an M.D. in 1900 and a Ph.D. in botany in 1903 working on myxobacteria. In 1907, he shifted his interests to genetics. Like most geneticists of his era (between the two world wars), he was sympathetic to and active in the eugenics movement, which had an international appeal. His main contribution to the text was the section on fundamentals of classical genetics. He was particularly adept in that role because he edited a series of monographs on genetics. Fischer and Lenz provided human genetics fundamentals such as monogenic inheritance of traits showing dominant, recessive, or X-linked inheritance; pedigree analysis; and polygenic traits (skin color, height, and other quantitative or variable traits). They also introduced eugenic and racial views and supported Ploetz's ideal of racial hygiene. It was this amalgam of pure science (Mendelian traits, chromosomal transmission, population genetics) and applied science (racial and social traits alleged to have genetic underpinnings) that appealed to Hitler and Nazi ideology. They wanted their beliefs in Aryan supremacy, racial hierarchies of abilities, and the existence of unfit people and inferior races to have a sound biological basis. German geneticists endorsed that view without having been National Socialists. They were motivated by different factors. Baur enjoyed his prestige as one of the leading geneticists in the world, whose work led to invitations to international conferences. He endorsed, or did not want to see, the views that Fischer and Lenz were promoting as a dubious science of racial ranking. His evil might be considered intentional neglect. He might also be accused of keeping bad company or sharing bias in co-authoring a text with so much bigotry in it (each edition got worse).

Eugen Fischer was less objective than he pretended to be. In 1942, he wrote with pride "...the results of the study of human heredity became absolutely indispensable as a basis for the important laws and regulations created by the new state."[8] Those Nuremberg marriage and sterilization laws of 1935 were founding documents of the racism and anti-Semitism characteristic of the Third Reich. What appealed to Fischer's vanity was that the science to legitimate those laws was drawn in large part from the work he and his colleagues had promoted. Fischer believed in racial differences based on heredity. He believed German Jews were a different people from Aryans. He liked the idea of science serving the state and the scientist playing the role of a behind-the-scenes inspirer of state policy. Fischer was also a bigot, although he denied this was so: "...the morals and actions of the Bolshevist Jews, bear witness to such a monstrous mentality that we can only speak of inferiority and beings of another species."[9] The language of science disguises (very poorly) the venomous racial ideology Fischer uttered in 1941. Fischer tried to vindicate himself after the war by arguing that his anti-Semitism was not virulent and he had considered the Jews an intelligent people with pathological social behavior. Because he did not condemn the Jews as intellectual inferiors, more virulent anti-Semites accused him of being a Jew-lover. He pretended (or deceived himself into believing) that throughout his career he was an objective scientist, a view his daughter also held when interviewed by Müller-Hill. Fischer's fawning personality is revealed in a letter written to von Verschuer in 1945 as Germany was rapidly losing the war. He began writing a text, *The Biological Basis of Racial Politics* "that would be non-political and for all races (Japanese or even Jewish)."[10] As Müller-Hill evaluated this ploy, "He was back at his old task, giving advice to rulers."[11]

Fritz Lenz was a patriotic German who suffered the humiliation of defeat in World War I and sought reasons for Germany's turmoil after the Armistice. He believed that biological forces were ruining the German race and that National Socialism would rescue it. He believed that eugenically unfit Germans were degrading the health, intelligence, and culture of Germany and that racial intermixture and marriages with Jews were changing the temperament of German people.

Lenz saw himself as a pure scientist who was eager to give objective scientific information to the state. He served on many committees and attended many conferences during the Third Reich years to supply evidence for the racial laws that were used to sterilize African-German hybrids (Rehoboth Bastarde), limit the activities of Jews living in Germany, assess or classify children with birth defects or mentally defective adults for euthanasia, and determine who is a Jew or a Gypsy. He accepted the view that Jews were a parasitic race (but argued that good parasites do not kill their hosts and thus the "disease" was not as severe as a cancer). He rationalized euthanasia by stating, "The life of a patient, who otherwise would need lifelong care, may be ended by medical measures of which he remains unaware." He saw euthanasia as a humane act, not a eugenic act.[12] He was a consultant on the resettlement of the eastern conquered territories and argued for biological-based standards: "The momentous events of the present time have enabled us to address the question of how we should recognize the racial quality of an individual and his racial value to our people."[13] Classifying Poles and Russians as Nordic-like or Slavic-like meant the difference between repatriation in a postwar society as good citizens (Nordic) or slave laborers (Slavic). Although Lenz may not have known Himmler's ultimate plans, he was certainly willing to help make the criteria for classifications for the use of the state. His son, Widukind Lenz, also a geneticist, saw his father as a scientist caught up in his times, and he rejected the image of guilt Müller-Hill assigned to his father: "A person standing on the outside to whom this whole period and the individuals aren't known through personal experience should be very careful how he judges it." He felt that Fischer, Lenz, and von Verschuer "were children of their culture, their time, and their class."[14] This is a classic rationalization to absolve all crimes as bad outcomes of environmental circumstances, and thus no personal responsibility can be assigned if the fault is the environment's.

Otmar von Verschuer had a distinguished career in anthropology and favored twin studies. Hence his connection with his research associate, Adolf Mengele, is particularly revealing about denial and guilt in a research scientist. We are not talking about a minor character in Third Reich academic circles. von Verschuer was head of a Kaiser Wilhelm Institute in Berlin during the Holocaust years. He, too, considered himself an objective scientist who only provided the state expert information based on objective standards. He helped set up the criteria for determining who is a Jew based on anthropological data, and he did a considerable number of classifications for "eugenic courts" that made inquiries into alleged half-Jews. He once complained to the Reich Minister of Justice that his testimony on the Jewishness of a wealthy half-Jew that he and Mengele had described as showing Jewish traits was overruled or disregarded by the court. According to his son, his father was strictly objective as a scientist and did not consider himself political. von Verschuer used this episode as evidence to both his Nazi critics and the Allies after the war that he was "incorruptible." For von Verschuer the mantle of scientific independence justified his cooperation with the government. He denied knowing that the eyes supplied to him by Mengele were from murder victims and claimed he never asked the source of his specimens, which included some suspicious data such as entire families with the same date of death. Like most of the theoreticians who were involved in Third Reich human biology, von Verschuer's correspondence, research data, and reports were destroyed or disappeared as the war ended. His son spoke protectively of his father; "I only know from my own experience that the motives which underlie past events are on occasion very difficult to fathom, and that even the opinions of those who were present at the time of the events aren't necessarily reliable."[15] This "turbidifying" of the record, making it too complex to tease apart, is another characteristic of the denial of evil.

Ernst Rüdin was director of the Psychiatric Department at the Kaiser Wilhelm Institute in Berlin. He was an ardent supporter of the race hygiene program and welcomed the National Socialist takeover in Germany, although he later denied being anything more than a theoretician and objective scientist. His daughter told Müller-Hill that her father only wanted those with severe genetic defects, not psychotics, sterilized under the Nüremberg laws, that he never was an anti-Semite, and that he was a scientist and not a politician.[16] Yet in 1943, Rüdin was enthusiastic in his praise for National Socialism: "Nevertheless, it will always remain the undying historic achievement of Adolf Hitler and his followers that they dared to take the first trailblazing and decisive steps towards such brilliant race-hygiene in and for the German people."[17] He criticized "parasitic alien races" (i.e., Jews and Gypsies) and "inferior stock" that needed eugenic pruning. If Rüdin had private reservations about the abuses of the race hygiene movement, he kept these largely to himself during the 13 years of the Third Reich.

The Foot Soldiers of the Holocaust

Very different were the motivations, careers, activities, and outcomes of those who did the actual selecting of who would be put to death and who would survive. Most of them were physicians (or trained health workers). As is often the case for war crimes (or other major crimes), the two most likely to be penalized are those on the very top and the foot soldiers who do the "dirty work." The middle management, the theoreticians, and the cheerleaders, who encourage or supervise these criminals, rarely suffer criminal prosecution.

Psychiatrist Robert Jay Lifton studied those who served in the hospitals and concentration camps where the killings were done. In both the euthanasia programs for the Aryan unfit and the "special handling," as it was coded and named after 1942 when the Final Solution was put into effect, the participant physicians did not consider themselves monstrous or even violating the values of their profession. Lifton concluded,

"...[W]hat I have struggled with throughout this study—is the disturbing psychological truth that participation in mass murder need not require emotions as extreme or demonic as would seem appropriate for such a malignant project. Or to put the matter another way, ordinary people can commit demonic acts."[18]

Just as Müller-Hill had a progression or escalation that reflected the transition from lesser to greater evil in the formation of Nazi ideology, Lifton had a progression for committing demonic acts:

- Forcible sterilization

- Medical euthanasia

- Nazification of the medical profession

- Concentration camp selection

- Mass murder

- Use of human subjects without consent[19]

Many of the physicians involved in the killings were idealistic and admired Hitler's creation of the Hitler Youth, whose camp meetings looked like innocent Boy Scout jamborees in propaganda films (e.g., Leni Riefenstahl's *Triumph of the Will*), and his social programs to provide good health to the vast majority of the German people. In a totalitarian state that provides employment, good health, and other social services to its people, there is reduced likelihood of young believers seeing flaws in its leadership. Some rationalized that in times of war when enemies kill healthy people, why should they object to releasing medical personnel to help those healthy soldiers and civilians in need instead of helping those whose lives were physically or mentally hopeless? Euthanasia of psychotics, children with birth defects, and senescent and retarded adults became acceptable, a lesser of two evils for those who felt uncomfortable with killing them and a civic duty to those who felt moral prohibitions against killing did not apply to people whose conditions forfeited the continuation of "worthless" life.[20]

Lifton describes many of the physicians whom he had interviewed as seeing their work as a "job to do." A small number who were bothered by their consciences sought to be transferred, and few were punished for that request. Another group performed their duties but drank a lot to compensate for their feelings. The combination of what today is called "post-traumatic stress syndrome" and alcoholism led many concentration camp directors to put the most unpleasant work (deceiving, leading, and pushing people into gas chambers; removing bodies for the crematoria) into the hands of Sonderkommandos—Jews and other inmates who got better rations for doing this work.[21] Their acts of evil the Sonderkommandos justified as doing what one has to do to save one's own life. One of the physicians described the adjustment period as, "In the beginning it was almost impossible. Afterward it became almost routine. That's the only way to put it."[22] Lifton calls this process "psychic numbing."[23] It is not unlike the breakdown of moral feeling described by Colin Trumbull among the Ik, a tribe forced to cease its hunter-gathering route after the political borders of its African nations set up barriers, leading to mass starvation and hopelessness.[24]

Medical experiments on inmates involved only a minute fraction of those killed or housed in the concentration camps. There were three types of experiments. Most notorious were those carried out on twins and those with birth defects by Adolf Mengele at Auschwitz. The results were forwarded to von Verschuer. The disappearance of those notes prevents a complete explanation of what was being studied. Presumably, the twin studies included controls and injected twins for pharmacological treatment of typhus and other diseases that inmates and soldiers were likely to encounter during stressful conditions. They also included twin studies for responses to toxins, burns, and other damaging agents. It is not clear what the intent was for the sets of eyes Mengele sent to von Verschuer. A second group of experiments involved experiments in sterilization, especially by X rays or (for females) formalin or phenol injections in the oviducts. Additional experiments included immersion in ice water to study survival and treatment for hypothermia. In a few cases, persons with particularly unusual deformities were selected upon decampment, and the bodies of these individuals were converted into skeletons to be sent to medical schools for study.[25] The justification used for these experiments was that these were people who would have been put to death anyway, and it was better to use their bodies for experimentation to help others or to provide science with knowledge that might not be available by other means.

Medical Experiments on Concentration Camp Inmates

Concentration camps in Germany and other countries were established to hold political prisoners, Jews, and potential enemies of the state. After the Wannsee Conference of 1942, many of these camps were assigned a role for the "final solution"—the extermination of the inmates by working them to death, by neglect leading to malnutrition and death, and by direct killing upon arrival or selection by prison staff. The most effective means of killing was by gassing (using cyanide) those arrivals and those selected in the camps as too feeble to work. Although this was the chief function of the death camps, some were selected to set aside a number of inmates and buildings for using inmates as human guinea pigs for medical research. This was true for Auschwitz, Dachau, Buchenwald, Ravensbrück, Sachsenhausen, Natzweiler, and Stutthof.

The most notorious of these medical experiments were carried out at Auschwitz, known to one of its camp physicians as the "anus mundi" for its horrific effects on the psyches of the Nazi staff.[26] German drug companies (I.G. Farben, Hochst-Leverkusen, Baeyer) asked permission to try out new chemotherapeutic agents on typhus and other infections of concern to the German military. Two products were used, ruthenal (an arsenical compound) and 3582 (a nitro-acridine).[27] Physicians were instructed to study the toxicity of these compounds as well as their efficacy against a variety of diseases. There were six physicians at Auschwitz on the medical staff (as well as Jewish doctors recruited to administer drugs and perform autopsies or other procedures for the staff). The inmates receiving the treatments had typhus (or were deliberately infected with typhus), tuberculosis, diarrhea, scarlet fever, or other infectious diseases. Other patients were infected to serve as controls, including those infected with typhus and not treated. Many of the patients claimed the treatments made them sicker. Jewish doctors tried to help the patients by getting them to spit out the drugs after they left the treatment facility or falsified records claiming that patients were treated when they actually had not received the drugs.[28]

A second project involved a comparative study of sterilization. Patients were sterilized by vasectomies, tubal ligation, castration of testes or ovaries, administration of X rays, or injections of chemicals into the reproductive tract. Some physicians, like Victor Brack, urged Hitler to sterilize the Jews instead of killing them, but Hitler rejected this option. Brack had been active in the euthanasia program that was initiated by Hitler before the Wannsee Conference. Some 200,000 children and adults who were retarded, psychotic, or who had severe birth defects were killed in a facility at Würtemberg. Brack used the pseudonym "Dr. Klein" on his death orders. He was later active in selecting arrivals he classified as "incurables" at the death camps for immediate gassing.[29]

Himmler thought sterilization of the Slavs and their use as slave labor would be very much desired after the war ended, with the Germans in control of eastern Europe. Himmler hoped to repopulate the occupied territories with an expanded Aryan race that he called the Lebensborn project.[30] These were Aryan children born to SS members and their Nordic women, some recruited from Norway, the Netherlands, and the Baltic territories, and some from Hitler youth groups for German girls. He approved Brack's experiments in sterilization at Auschwitz. Males were administered radiation doses of 500–600 r and females a dose of 300–350 r. The participating doctors reported radiation burns and abscesses in the inmates. Testes or ovaries of those irradiated were surgically removed and analyzed for the radiation effects on the germinal material. Many of those sterilized were sent to the gas chambers as too sick to work after their radiation exposure and surgery. Those female inmates sterilized by chemicals were usually administered phenol.[31]

One physician (Carl Clauberg) wrote to Himmler, "The method invented by me of sterilizing the female body without surgery is already completed. It consists in a single injection into the cervix uteri, which can be practically performed by one physician who knows something about gynecological examination. If I maintain that the method 'is almost ready', I meant: 1/ that some small perfections have still to be made, 2/ It could already be introduced today during our normal surgical sterilizations instead of surgery and could replace the latter. ... One properly trained physician, in a properly prepared place, with the help of more or less ten helpers (the number of helpers depends on the demand for acceleration) could probably sterilize several hundred but not over 1000 a day."[32] Clauberg's injection of phenol into the oviducts was extremely painful. He was a respected gynecologist before the start of World War II and had published articles on the corpus luteum in reputable medical journals. He was a professor at Königsberg.

Assessment of War Crimes by International Tribunals

When the war ended and Nazi leaders were captured, many were charged with war crimes. The historical story of war crimes begins with Napoleon's exile to Elba and his escape. He was named an "international outlaw," and after his defeat at Waterloo, he spent the rest of his life exiled in St. Helena by "executive action." In World War I, war crimes were again raised by the Allies in 1919, but the U.S. opposed the concept because it opposed international courts and did not want to innovate such a resource that could be used against its future policies and citizens.

Matters changed in World War II. Exiles from occupied countries revealed numerous atrocities against civilian populations. President Roosevelt and Prime Minister Churchill in 1942 agreed to prosecute Nazi war criminals when the war ended. Henry Stimson recommended an "international tribunal" to prosecute them and President Truman recommended the formation of such a tribunal when the United Nations was being planned. Truman appointed Robert H. Jackson to prosecute the Nazi and Japanese war criminals. Jackson proposed "to punish acts which have been regarded as criminal since the time of Cain and have been so written in every civilized code."[33]

There were eleven acts of war crimes, including conspiracy, murder of civilians, deportation for slave labor, murder of prisoners of war, killing hostages, plunder of property, extraction of collective penalties, wanton destruction of cities, Germanization of occupied territories, use of concentration camps for killing, and religious, racial, or political mass killing.

There were 21 top leaders of the Third Reich who were prosecuted by Jackson, who called them "an evil alliance." With the exception of Hermann Goering, all the indicted Nazis blamed Hitler or those superior to them for issuing orders they felt compelled to obey out of duty, patriotism, or the law.[34] Goering justified the alleged atrocities as attributes of modern war. The use of air power and powerful weapons made civilian casualties inevitable, and modern war was thus "total war" in which an entire nation was pitted against another for survival; thus, the usual rules of war did not apply. This included the right to appropriate possessions and territory and to eliminate both real and potential enemies. Goering also challenged the legality of the International Tribunal because there was no precedent for it.

At the Nuremberg trials, Jackson introduced the concept of "crimes against humanity," a phrase introduced at the 1907 Hague Convention and used by its membership in 1919 to criticize the Turkish government for its massacre and expulsion of Armenians in that country. The term "evil" was rarely used during the Nuremberg trials. Phrases such as "policy of corruption," "war of fanatical religion," or "horrible crimes" were used instead.[35] Only a small portion of the charges against the top Nazi leadership was related to the Holocaust. Jews were not used to present evidence in order to defuse any charges by the Nazi defense that this was a vendetta by Jews. Nevertheless, Jackson did use the term evil to describe the plight of Jews: "The Nazi movement will be of evil memory in history because of its persecution of the Jews, the most far-flung and terrible racial persecution of all time."[36]

There were 12 additional Nuremberg trials for other Nazi leaders of second rank (about 185 defendants). There were additional trials (many thousands) in individual countries and in Germany for war crimes, betrayal, and collaboration. The Cold War diminished interest in additional prosecutions after 1949 as the Allies in the West aligned themselves against the Soviet bloc and needed Germany as a buffer against Soviet expansion into Europe. Included among the 12 Nuremberg trials were the medical experiments headed by Karl Brandt and his chief physicians at concentration camps where involuntary use of inmates was documented.[37] Only a small portion of the 200 million pages of evidence was used to convict the top designers and implementers of these medical experiments on inmates. Six received death sentences and three had committed suicide rather than face trial.[38] Many disappeared by changing names or escaping to other countries to practice medicine.

Assessing Evil Acts in Nazi Biology and Medicine

It is hard to pretend that the compulsory sterilizations, euthanasia, mass murders, and experimentations on human subjects are not acts of evil. Had those Nazis involved asked the question Immanuel Kant raised for an ethics based on reason, they would have asked, "Would I want myself to be sterilized, killed, or experimented on without my consent?" and in all likelihood they would have answered no. That is the answer of a rational person. If all rational people felt likewise, then Kant would consider the response to be universal and such behavior unethical. They did not ask themselves that question because they equated their victims as subhuman. They removed humanity from their victims by seeing them as parasites, degenerates, mentally unfit, physically deformed, or behaviorally corrupt. These definitions of eugenically unfit individuals allowed them to rationalize that they were killing enemies, preventing future infections, purifying the Volk, or acting in a humane way.

People consider themselves humane when they shoot horses with broken legs, when they bring sick pets to a humane society to be injected with a lethal agent, or when they administer a "coup de grace" for a person still twitching or moaning after being shot by a firing squad. There may have been some SS doctors who actually were sadistic killers and bigots. Most were not, but they went ahead anyway doing their unpleasant tasks with a full set of rationalizations from the law, from their superiors, and from their country. Although the comparison may be made (as Lifton later did) that those who dropped atomic bombs on Japan in 1945 were similar to SS doctors who participated in the Holocaust, it is much more difficult for most Americans to make such an identification. As we shall see when we look at the Japanese-American war in the Pacific and the Cold War and its effects on scientific thinking, there may be some (but not a compelling) truth to Lifton's insight. What distinguishes Nazi medical and life science is the magnitude of the evil, the virtually complete complicity of the Nazi doctors in the objectives of race hygiene, and the desire by a state to kill entire peoples (not isolated targets). Whatever the motivations might have been of the U.S. flight crews, or of those who sent them, they were not intent on a Final Solution for all Japanese.

Notes and References

1. For a brief account of race hygiene and its role in the formation of Nazi ideology, see Elof Axel Carlson, *The Unfit: A History of a Bad Idea*, Chapter 18, "The Smoke of Auschwitz," pp. 315–335 (Cold Spring Harbor Laboratory Press, Cold Spring Harbor, New York, 2001).
2. For a thorough account of Hitler's career, see Michael Burleigh, *The Third Reich: A New History* (Hill and Wang [A division of Farrar, Straus and Giroux], New York, 2000).
3. Herbert Spencer, *Social Statics: The Conditions Essential to Human Happiness Specified, and the First of Them Developed* (Chapman, London, 1851). Spencer's book is the founding document of the Libertarian political movement.
4. Robert Jay Lifton, *The Nazi Doctors* (Basic Books, New York, 1986).
5. Benno Müller-Hill, *Murderous Science: Elimination by Scientific Selection of Jews, Gypsies and Others, Germany 1933–1945* (Oxford University Press, New York, 1988). (There is also an expanded edition of this book: Cold Spring Harbor Laboratory Press, Cold Spring Harbor, New York, 1998.)
6. Ibid., p. 5.
7. Ibid., p. 23.
8. Ibid., p. 61.
9. Ibid., p. 46.
10. Ibid., p. 80.
11. Ibid., p. 80.
12. Ibid., p. 40.
13. Ibid., p. 77.

14. Ibid., p. 115.
15. Ibid., p. 118.
16. Ibid., p. 123.
17. Ibid., p. 61.
18. Lifton, op. cit., p. 5.
19. Ibid., p. 5.
20. Ibid., p. 112.
21. Ibid., p. 252.
22. Ibid., p. 15.
23. Ibid., p. 195.
24. Colin Trumbull, *The Mountain People* (Simon and Schuster, New York, 1972).
25. Lifton, op.cit., p. 285.
26. International Auschwitz Committee, *Nazi Medicine: Doctors, Victims, and Medicine in Auschwitz* (Howard Fertig, New York, 1986), p. 1.
27. Ibid., p. 32.
28. Ibid., p. 19.
29. Ibid., p. 50.
30. Catrine Clay and Michael Leapman, *Master Race: The Lebensborn Experiment in Nazi Germany* (Hodder and Stoughton, London, 1995).
31. International Auschwitz Committee, op. cit., p. 68.
32. Ibid., p. 75.
33. Michael R. Marrus, ed. *The Nuremberg War Crimes Trial 1945–46: A Documentary History* (Bedford/ St. Martin's Press, Boston, 1997), p. 43.
34. Ibid., p. 179.
35. Ibid., p. 191.
36. Ibid., p. 193.
37. Donald Bloxham, *Genocide on Trial: War Crimes and the Formation of Holocaust Theory and Memory* (Oxford University Press, Cambridge, 2001), p. 231.
38. Bruce M. Stave and Michele Palmer with Leslie Frank, *Witness to Nuremberg: An Oral History; American Participants at the War Crime Trials* (Twayne Publishers, New York, 1998), p. 3.

4

The Banality of Evil

The Careers of Charles Davenport and Harry Laughlin

H ANNAH ARENDT, IN *Eichmann in Jerusalem*, made a controversial assessment of Adolph Eichmann by describing his career as "the banality of evil."[1] Eichmann was the bureaucrat Nazi who scheduled the deportation of Jews to the killing centers in eastern Europe, and he was the recording secretary of the Wannsee Conference in 1942 that designed the "final solution" or liquidation of Jews from occupied Europe. Many objected to Arendt's view that ordinary people are corruptible by self-interest or shallow values and capable of participating in mass murder. Those who objected preferred to see a character flaw or capacity for evil in Eichmann that is exceptional in human behavior. The debate is important, because if acts of mass murder are pathological, then only a small portion of humanity has this demonic trait, whether innate or culturally created. If Arendt is right, we have much more to worry about because humanity is constantly vulnerable to circumstances that can make ordinary people participate willingly in extraordinary crimes against humanity. At the same time, Arendt's idea of the banality of evil gives humanity an opportunity to find ways to educate its future generations to prevent such corrupting influences. I share with Arendt this latter view and consider the two chief promoters of the American eugenics movement, Charles Benedict Davenport and Harry Hamilton Laughlin, to be prototypes of Eichmann.[2]

Davenport's Career as a Geneticist

Davenport's career included his work as a geneticist, his interests in eugenics, his desire to educate others about the life sciences, and his administrative skill in building and directing a major center where these three activities could be carried out. Davenport was educated in the late 19th century. He took an early interest in natural history because his father, a real estate broker in New York City, had a farm in Connecticut where they spent the summers. Davenport's father was strict and wanted a secure future for his talented son. He felt farming and natural history were not profitable careers and pushed him into engineering instead. Davenport followed his father's wishes but kept up his interests in evolution and experimental science. Eventually his father relented and Davenport returned to college for a Ph.D. in zoology. He made a name for himself by publishing a two-volume work on experimental morphology as he saw it in the late 1890s.[3] It was very quantitative and strongly influenced by the prevailing Darwinian model of evolution then taught in the major universities. Character traits in this model changed very gradually, if not imperceptibly, from generation to generation. Scientists would look for subtle changes in response to temperature, centrifugation, chemical stimuli, and other environmentally introduced agents in controlled experiments. Morphology was used as the way to measure such changes with careful measurements of the lengths and ratios of limbs, wings, mouthparts, or other minor details of features.

In 1900, this prevailing biometric model championed by W. F. R. Weldon and K. Pearson was upended by the discovery of Mendel's long-forgotten findings of 1865 which showed that many traits were discontinuous and their distribution across generations was highly predictable through breeding analysis. Davenport immediately recognized the importance of the new Mendelism, and he contributed to it by showing that a number of factors in chickens were inherited according to Mendelian laws.

Davenport, independently of William Bateson in Great Britain, demonstrated that the various combs in chickens were a consequence of the interaction of two different genes (and in some cases of three different genes).[4] He worked these out and published those results while still at Harvard. He argued that the discovery of typical and atypical Mendelian traits in animals supported a universal model of hereditary units determining character traits. He was an enthusiastic supporter of Bateson's work, which provided the most widespread confirmation and extension of Mendelism in the scientific world.[5] Davenport was an enthusiast for the new

Mendelism and urged his colleagues in natural history, zoology, and botany to embrace it as a new tool for studying evolution.

Davenport was seen as a multi-talented, ambitious, and rising star among the older members of the American Breeder's Association. He also knew how to circulate among the wealthy elite, whose children took his courses at the Brooklyn Academy. He was selected to be fund-raiser, designer, and administrator for an enterprise set up on the north shore of Long Island, New York, at Cold Spring Harbor. Eventually, this housed all three components of Davenport's interests: the Long Island Biological Station provided the summer courses for high school teachers; the Carnegie Institution of Washington supported the basic genetic research of his permanent staff and the summer investigators who came to do uninterrupted research; and the Eugenics Record Office carried out the basic and applied human genetics that fed into the growing American eugenics movement encouraged by the American Breeder's Association and funded by a gift from the Harriman estate. The Long Island Biological Station was separately funded by donations from the wealthy families that lived on the north shore of Long Island. The building of this empire of scientific activities took place between 1903 and 1913. Davenport proved effective in designing the facilities and creating an atmosphere where basic research could be carried out by its permanent and visiting investigators, allowing scientists to exchange ideas and enjoy the stimulation of their colleagues from other universities. This was not original to Davenport; he used the Naples Station for Marine Biology in Italy as a model, even to adopting its architectural design for its chief buildings. He was also inspired by the Wood's Hole model in Massachusetts, which had a powerful impact on experimental zoology and developmental biology. The Cold Spring Harbor Laboratory was devoted to genetics and experimental evolution.[6]

Davenport's Personality

Davenport was insecure. This is a feature of many eminent persons. They compensate for their insecurities by driving themselves to prove their worth. But Davenport was also autocratic. He ran his own show and liked to surround himself with weak assistants who admired him and who were unlikely to take issue with his major views. He liked to hear good things about his work and was defensive and intolerant of criticism. These character flaws made him vulnerable to error, bias, and the corrupting influence of power.

His childhood need to prove himself to his father also played a role in his career. He saw himself as a potential benefactor to the world through his eugenic movement. In Davenport's career, these led to an over-inflated view of his contributions as a geneticist, as leader of the eugenics movement, and as a power broker in society. Davenport does not rank with E. B. Wilson, T.H. Morgan, or R.A. Emerson among the first rank of the new Mendelians in the U.S. who dominated classical genetics in the first decades of the 20th century. Nor did he catch up to them in discoveries of major importance during the years of his work at Cold Spring Harbor. This is understandable after 1910 because he was heavily involved in administration and could not do his own research on a large scale. Davenport was nevertheless recognized as a capable geneticist and a solid scientist who had earned his reputation as a leader in that field. He was elected to the National Academy of Sciences and certainly had the publication record and the quality of work from his early years to merit this distinction. Nor should one doubt the effectiveness of the Cold Spring Harbor Laboratory in promoting good science through the Carnegie Institution's supported research. The publications of its basic genetic research set a standard of excellence that has continued throughout the 20th and into the 21st centuries.[7]

Davenport's Contributions to Human Genetics

It was in human genetics that Davenport sought his greatest fame. If he could analyze human traits and show which ones Mendelized and which did not, he could identify traits of value to medicine, psychology, and society. He recognized that humans cannot be bred like chickens, mice, or fruit flies to satisfy a scientist's curiosity, and he chose pedigree analysis as the most effective means of carrying out his analysis.

Pedigree analysis was initiated by Francis Galton but not for Mendelian traits. Galton used this method for what he thought were quantitative traits that ran in families—especially behavioral traits such as unusual talents in music, mathematics, and writing.[8] It was a way of showing the preponderance of family members with eminent or potentially eminent traits. After the Mendelian rediscovery in 1900, Davenport was one of the first to apply pedigree analysis to human physical traits. He sought the help of physicians and hoped to train them in taking accurate pedigrees. This was part of his enthusiasm for a Eugenics Record Office, where such pedigrees could be maintained and where he could train field workers to visit institutions and take family histories of the patients. It was inspired also by the work of Richard Dugdale, who compiled family histories of criminals and paupers among kindred of Dutch settlers from the early 1700s who had proliferated in the Hudson Valley near Kingston (Ulster County in New York). Dugdale was an environmentalist and his popular book on the Jukes, as he called this family, became a classic of sociology in the late 19th century.[9] Those who read it rejected his interpretation that a good environment could reverse the bad behavior of the Jukes, and they embraced his data, reinterpreting it (from the 1880s on) as a more pessimistic evidence for fixed defects of protoplasm that were being passed on, corrupting future generations. Davenport was one of the strongest supporters of this hereditarian interpretation of the Jukes kindred. His education was heavily influenced by August Weismann's theory of the germ plasm, which (correctly) identified a separation of environmental influence on the soma (the body cells) and a relative isolation of such environmental effects on the germinal cells of the gonads (the germ plasm).

Davenport demonstrated that albinism in humans is an autosomal recessive trait (one not associated with a sexual difference in incidence).[10] The parents of such a child are carriers (heterozygous in the geneticist's technical jargon) and normal in appearance, but each of their reproductive efforts leading to a child has a 25% chance of bringing the recessive genes together and producing an albino child. He also demonstrated that human skin color is a quantitative trait. He went to Jamaica in 1912 to do an extensive study of children and grandchildren of parents who were interracial couples.[11] He and his coworkers used a color wheel resembling a spinning whirligig that had different sectors of white, black, red, and yellow to produce a blended blur that they would apply to the inner arm of the subjects. When a matching color was found they would have a quantitative measure of the color rather than a subjective term to describe it. Davenport's findings were impressive. He exploded many folk myths about human racial hybridizing. He claimed there were two chief genes involved in human skin color and their effects were additive. There was no dominant or recessive factor for color. A person expressed as many of the color factors as were present in the genetically inferred composition (genotype) of that individual. If the color factors are A and B for melanizing pigment and a and b for the virtual absence of melanin in the skin, then African males or females are AABB and white males and females (especially the Anglican whites in Jamaica) are aabb. Their children who have brown skin are AaBb. When two such brown-skinned individuals have children, their offspring form a spectrum of colors in a fixed ratio of 1 AABB: 4 AABb or AaBB: 6 AaBb or AAbb or aaBB: 4Aabb or aaBb: 1 aabb. Converting the intense melanin-producing factors into a color effect, this would be seen as 1 black: 4 dark brown: 6 brown: 4 light brown: 1 white. Davenport rejected a myth in southern bigotry that a white person with a black ancestor later marrying a white woman with no known black ancestry could have a black baby. His work also explained why two light brown-skinned parents could have a child darker than either parent. Thus, Aabb (light brown) x aaBb (light brown) can give one-fourth of the offspring having AaBb (brown), one-fourth aabb (white), and half Aabb or aaBb (light brown). Davenport used the work of H. Nilsson-Ehle (working with cereal grains) and E. M. East (working with maize) for this model that he applied to human skin color as a quantitative trait.[12]

Davenport was also the first to recognize and interpret what is called a founder effect in human genetics. He noted that families which are isolated geographically, socially, or by religion become genetic isolates and marry with one another. He identified each such isolated population running along the Atlantic coast from Maine to Virginia with unique Mendelian defects that were accidental manifestations of what was brought into that population by a carrier ancestor.[13] He identified deaf mutism, albinism, midget stature, and similar recessive traits expressed in these different communities. Sometimes a dominant trait (such as Huntington disease) could be found in a community where people shunned marrying into their families and tended to breed among themselves, as the original Dr. George Huntington described on the south fork of eastern Long Island when he first described this disorder.

Davenport and the American Eugenics Movement

Davenport adopted a form of eugenics that is called negative eugenics by historians of science. Galton, who coined the term eugenics in 1883, founded a movement that is more properly called positive eugenics. In Galton's idealistic view, the history of a civilization could be measured by the contributions of a few eminent individuals. We think of Pericles, Aristotle, Plato, Socrates, Solon, Euripides, Aeschylus, and Sophocles when we dredge our memories on what made the Golden Age of Greek civilization so golden. If we were to mine the 19th century for its major contributors, we would think of Napoleon, Lincoln, Marx, Darwin, Faraday, Pasteur,

Koch, Beethoven, Van Gogh, Tolstoy, Hugo, Wordsworth, and Goethe as a sampling of the many hundreds who could be singled out for their great contributions to that century. This is known as the heroic theory of history. It assumes that major contributions to science, art, politics, literature, religion, and other hallmarks of civilization are the products of a small percentage of eminent individuals whose names we revere. Galton believed they are a national treasure who should be encouraged to reproduce more frequently than ordinary people because of their special talents. He tried to prove that such traits were inherited and claimed in his books on *Hereditary Genius* and *Natural Intelligence* that about 20% of the children or parents of an eminent individual were themselves eminent.[14] Galton believed eugenics was a means of increasing the pool of talent (and thus the benefits to humanity) if they had many more children than they normally would produce. Historians call Galton's views positive eugenics.

Negative eugenics makes a different assumption. It assumes that the basic stock of a nation is healthy, and opportunity will distinguish those with ambition and talent from those who lack these traits. But among the failures of society are some notorious populations of thieves, feeble-minded individuals, lunatics, beggars, and vagrants who corrupt society by their petty crimes, demands for welfare, and ill health, requiring hospitalization and the construction of jails to keep them from preying on otherwise decent people. In the U.S. in the last half of the 19th century, families such as the Jukes in New York state and the Tribe of Ishmael in Indiana were held up as examples of pathological social failure who harbored a defective germ plasm.[15] It was a theory of heredity similar to that of infectious diseases. A contaminating individual would corrupt the children of an innocent person who married such a defective person out of ignorance. Advocates of negative eugenics argued that society needed to protect itself by isolating the contaminating strains. This led to an expanded asylum movement (less for treatment than for storage to protect society) and eventually to more drastic measures such as restrictive marriage laws and compulsory sterilization laws.

These trends were already in place when Davenport received his college education and began making a name for himself as a geneticist. He shared a sympathy with those like David Starr Jordan and Alexander Graham Bell, who asked him to serve as a secretary to their newly formed committee on eugenics for the American Breeders Association. Jordan was a well-known ichthyologist and evolutionary biologist who was also a president of Indiana University and first president of Stanford University. He was a prolific writer and popularizer of science, with books on heredity, eugenics, and evolution as well as criticisms of war (he was a leading pacifist and friend of Jane Addams).[16] It was Jordan's essays that popularized the work of Oscar McCulloch on the Tribe of Ishmael. He denounced war as dysgenic, killing the most able and allowing those unfit for military service to stay home and reproduce. He also denounced the waging of war because it ignored a potential to establish an international court to adjudicate disputes among nations. His third argument against war was that it was costly and robbed nations of budgets that should be used for the education and health of its people.

Bell is best known to us as the inventor of the telephone. He had a long-standing interest in the teaching of the deaf and compiled evidence that some forms of deafness were inherited. He was concerned that sign language would encourage the deaf to form a race of their own, keeping them culturally and reproductively isolated. He also took an interest in the heredity of supernumerary breasts in sheep and demonstrated that these formed along the milk line; such extra breasts were frequently present in some strains of sheep.[17] Both Jordan and Bell were enthusiasts for the new field of genetics that was emerging, and they urged the addition of a third wing to the already robust plant and animal genetics of the American Breeder's Association when they proposed adding a eugenics committee.

Davenport may have felt that although he was good at genetics research, it was not his forte. By focusing on the administration of a science complex devoted to genetics, he would certainly gain respect and recognition for the staff he would recruit. Eugenics, however, was not like experimental genetics, with its heavy demand for field work, microscopy, or crosses of living things. Those were time-dependent and did not allow much interruption for administrative work. Human genetics was different. Davenport could study pedigrees in his leisure time. He could interrupt a study of a trait and go back to it without having to start all over. But there was something special about human genetics research that made it appealing. It gave to science the potential to redirect evolution, to compensate for civilization's more charitable effect on those with defective germ plasm. It could identify the problem families and the problem traits and give those families or society itself an opportunity to limit their reproduction.

People with good intentions do not have evil thoughts in their heads. They believe they are doing the right thing. What they lack is a Promethean foresight into the implications of what they are advocating. This is why, I believe, so much harm can be done by people with good intentions. They can lack the curiosity or talent to reflect on the possible outcomes of their ideas. Davenport can be faulted for more than this level of ignorance, which is shared by most of humanity. He was happy acquiring power, whether it was in his domain over the Cold Spring Harbor enterprise or for the emerging field of the American eugenics movement, which he tried to shape to his own values. It is this latter aspect which justifies my associating him with Arendt's judgment of practicing the banality of evil.

The Objectives of the American Eugenics Movement

The scope of Davenport's interests in eugenics is revealed in his 1911 book, *Heredity in Relation to Eugenics*.[18] He was shaped by 19th-century thinking about human heredity and the classification of human traits. Modern medicine, he claimed, "has forgotten the fundamental fact that all men are created unequal in their protoplasmic makeup and unequal in their powers and responsibilities."[19] His use of the term "protoplasmic" instead of "genetic" reflects this older perception. It is an unfortunate one, because protoplasmic implies a contaminating model, like an infectious disease that corrupts the child of a defective parent. The newer Mendelism should have signaled a different possibility of diverse outcomes from the breeding of allegedly defective human beings. Thus, albinos generally produce normally pigmented children when they have normally pigmented spouses. Persons with Huntington disease produce half their children without the potential for the disorder when they have children with a partner who has no history of the disorder.

Davenport's list of hereditary traits is an interesting compilation, "specifically, the Record Office sees pedigrees of families in which one or more of the following traits appears: short stature, tallness, corpulence, special talents in music, art, literature, mechanics, invention, and mathematics, rheumatism, multiple sclerosis, hereditary ataxias, Ménière's disease, chorea of all forms, eye defects of all forms, otosclerosis, peculiarities of hair, skin, and nails (especially red hair), albinism, harelip and cleft palate, peculiarities of the teeth, cancer, Thomsen's disease, hemophilia, exophthalmic goiter, diabetes, alkaptonuria, gout, peculiarities of the hands and feet and of other parts of the skeleton."[20] Note the absence of 19th-century categories of human heredity—criminality, psychosis, mental retardation, vagrancy, and pauperism. It is not that Davenport has abandoned these. He wants his list to reveal the potential of the new science of genetics to identify a hereditary basis for the "peculiarities" of human variation. He hoped to use this encyclopedic listing of potential hereditary traits as evidence that the "unsocial classes," as he called them, were indisputably hereditary in their origin. He also assumed, erroneously, that the individuals of the unsocial classes would not be born to the middle class and to the elite readers of his books. In an oft-quoted claim, he argued "We have become so used to crime, disease, and degeneracy that we take them as necessary evils. That they were so in the world's ignorance is granted; that they must remain so is denied."[21] Davenport's use of the phrase "necessary evils" and his emphatic denial that this must be so reflects a messianic personality. He wants to convince his readers that eugenics has a social role. This includes the right of the state to control the propagation of the mentally incompetent and a rational approach to marriage. But it also presented a dilemma. Although he opposed abortion or euthanasia (destruction of the unfit before or after birth), he was concerned about the new compulsory sterilization movement launched in the late 19th century by Harry Clay Sharp in Indiana. Indiana had become the first state to make such sterilizations of degenerates legal. Davenport worried that the sterile degenerates would contribute to promiscuity with no fear of having to rear unwanted children. Later, he would abandon that argument after his student Harry Laughlin convinced him that the consequences of asylums releasing fertile degenerates would be of even greater danger to society.

Harry Laughlin and the Eugenics Record Office

Harry Laughlin was a student in Davenport's summer program for high school teachers when it was still located in Brooklyn. Davenport encouraged Laughlin to get a Ph.D., and he contacted Edward Conklin at Princeton University. Laughlin did an undistinguished dissertation on mitosis in the onion (*Allium*) and came back with dual interests in studying the pedigrees of horses, especially thoroughbreds, and human heredity. Davenport convinced him to do the thoroughbred studies as a hobby and to devote his major efforts to running the Eugenics Record Office. He became the superintendent of the Office in 1913 and regularly attended regional, national, and international meetings on eugenics. Laughlin was born in Iowa and grew up in Missouri. His mother was a suffragette, and he admired her social activism. He had several older brothers who were successful osteopaths, and his father was a college president.[22] Laughlin preferred natural history, and that was what attracted him to Davenport's institute. Laughlin was not as sophisticated as Davenport nor as broadly educated. He adored Davenport, who lifted him from the obscurity of being a country teacher and gave him a position of responsibility and the credentials to be accepted by the elite. He lived in Davenport's shadow and wanted to impress him by finding a place in eugenics that would meet his limited talents. He found that in several outlets. He was an excellent bureaucrat, and he could amass immense detail and organize it and present it. He liked lobbying for eugenics, and he recognized two opportunities to do so, one in state sterilization laws and the other in restrictive immigration legislation. An additional interest was that of being a de facto legal scholar. He affiliated himself with legal scholars so he could prepare model eugenic laws that would meet the test of constitutional challenge at the state or federal level.[23]

Laughlin had few self-doubts about his values for eugenics. He believed inferior people were a menace and needed to be isolated, sterilized, or banned from entering the country. At home, the issues Davenport stressed were primarily home-grown paupers, psychotics, and the feeble-minded. But Laughlin saw an additional category emerging. He identified southern and eastern Europeans as the riff-raff of European countries who were dumping their problems on America's shore. Restrictive immigration legislation was the best response to this problem. Fortunately for Laughlin, his prejudices were widely supported in the U.S.[24] The new immigrants often became labor union organizers. They brought alien ideologies, especially socialism, to working-class Americans, corrupting them with a belief that their only salvation was the destruction of capitalist society. They spoke foreign languages and read their own foreign language newspapers printed in the U.S. Many lived in ghettoes. They introduced an organized crime with picturesque names (the Black Hand, the Mafia, or Murder, Incorporated). Their health was bad, and many were unable to find work or keep a steady job. The Italians, Balkan nation immigrants, eastern European Jews, Serbs and Slavs of all sorts, Middle East Muslims, and hordes of Catholics were a threat to a once predominantly Protestant country with largely British and western European descendants.

Laughlin's Approach to Eugenics

Laughlin began his first effort by assessing the eugenic sterilization laws. He considered most of them worthless, because they did not meet constitutional guarantees of due process and consistency. In 1914, he promoted Davenport's priorities of sexual segregation in asylums with sterilization only for those who would be released. He used phrases for the "unfit" such as the "submerged tenth" and "defective germ plasm," as carryovers from 19th-century perceptions of heredity. After his Ph.D., he shifted to a Mendelian vocabulary in describing those who should be sexually segregated, and his enthusiasm for compulsory sterilization greatly increased. By 1917, the Carnegie Institution was getting nervous, and they did not want Laughlin to lobby for sterilization laws.[25] They felt this would imperil their charitable status. They asked Davenport to assign Laughlin to other, less political, activities. Laughlin chose to serve as an "expert witness" at hearings in state and federal committees dealing with eugenic issues.

Laughlin did so because after 1912, a shift in eugenic interest took place. Prior to that year, the main concern was over the Jukes and other U.S.-born and rooted families. After 1912, Laughlin took a deep interest in the national origins of defectives in relief rolls, hospitals, prisons, and asylums. Within a decade, he established himself as a regular expert witness for the House Committee on Immigration and Naturalization, and he became a friend of Albert Johnson, its chair.[26] Johnson was a Republican representative from the state of Washington and virulently opposed to Japanese immigration to the West Coast. Laughlin struck up a friendship with Harry Olson, the Chief Justice of the Municipal Court of Chicago, and he worked on model eugenic laws with him and his staff.[27]

This eventually led to cooperation between the Eugenics Record Office, the state of Virginia, and both sides of what became known as the *Buck v Bell* Supreme Court case of 1927. In that case, the legal principals in Virginia were friends and mutually agreed that whoever lost in court would continue to appeal the case to the Supreme Court. In a brilliant analysis of this case, Paul Lombardo demonstrates the way Carrie Buck was selected and railroaded to her eventual sterilization, although it was clear from the evidence, even at that time, that Carrie Buck's imbecility was highly questionable.[28] The 8-1 decision upholding the right of Virginia to sterilize its unfit citizens was the high point of the efforts of the Eugenics Record Office to put a eugenics program into practice.

Laughlin's efforts in promoting eugenic sterilization and restrictive immigration legislation based on the alleged inferiority of southern and eastern Europeans was much admired by the growing racial hygiene movement in Weimar, Germany, and embraced by the Nazis when they came to power. Laughlin was thrilled and boasted, "To one versed in the history of eugenical sterilization in America, the text of the German statute reads almost like the 'American model sterilization law'."[29] Laughlin was honored in 1937 with an honorary doctoral degree from Heidelberg University for his contributions to eugenics, but the State Department advised him not to make the trip to Germany to accept the award. Instead he went to a smaller ceremony at the German consulate in Rockefeller Center in New York City.

The award was too late for Laughlin's career. The election of President Roosevelt in 1932 ended the Republican control of Congress, and Laughlin found he was no longer of interest as an expert witness. Davenport retired as head of the Cold Spring Harbor Laboratory in 1935, and an external committee evaluated the work of the Eugenics Record Office and found virtually all of its eugenic research flawed or unworthy of further support.[30] Laughlin resigned and went back to Missouri to live out his retirement years.

Assessing Davenport and Laughlin

As a geneticist, Charles Davenport was never as creative, successful, committed, or gifted as his American contemporaries Morgan, Wilson, Castle, Emerson, East, or Shull. They spent their entire lives as experimentalists and made substantial contributions to classical genetics. Davenport essentially dropped out as an experimentalist after 1910, when almost all his activities were directed to eugenics and the administration of a first-class facility for studying evolution and genetics. Unlike Galton, who was a brilliant theoretician, Davenport lacked an original mind. He compensated by being an effective administrator. In this respect, he was like Anton Dohrn, who founded the Naples station in the last quarter of the 19th century. That was also true for Fernandus Payne, who became a successful Chair and Dean at Indiana University and who built a great program in genetics. Davenport was too ambitious to allow his fame to be that of an able administrator. He believed he was as good as his more famous contemporaries, and he used his power to serve on many committees promoting genetics. His good intentions were always tinged with an ambition to gain recognition. He shared many of the social prejudices of those philanthropists with whom he felt at ease. He had an opportunity at the Eugenics Record Office to develop a Mendelian study of human traits, but he was so convinced that behavioral traits were the key to this effort that he tried to force into that new Mendelism the "submerged tenth," the unfit, and the traditional social classes of criminals, psychotics, feeble-minded, paupers, and vagrants.

Laughlin had the misfortune to be worse. He knew he wasn't in the same league as the geneticists whose articles filled the pages of the journal *Genetics*. His work on thoroughbred horses was never successful in identifying any genes that made great horses, and the vast amount of his research went undigested and unpublished. His eugenics research was similarly void of useful publications. He satisfied himself with the only outlet for a person lacking research sophistication or creativity. Although Laughlin was never a rabid anti-Semite nor an overt bigot, as many of the more extreme officers of the KKK were during that era, he revealed his views in a letter to Madison Grant (a notorious anti-Semite, white supremacist, and racist writer).[31] In 1932, he differed with Hitler's extreme position on Jews, but he told Grant, "a Jew must be assimilated or deported."[32] As Davenport's bureaucrat, Laughlin took the heat for many of the controversial political forays they made to extend eugenics to society. Unlike Galton, who saw eugenics in the late 19th century as a moral effort to educate his fellow elitists to have more children, Laughlin saw eugenics as a political effort to identify unfit classes and individuals and isolate them from reproducing.

But the work of Davenport and Laughlin did lead to the sterilization of over 40,000 Americans, and they lent their moral support to the early years of the race hygiene movement and the Nazi Nuremberg laws. Had the Nazis won the war, I do not doubt that if Davenport and Laughlin had been alive and in good health, they would have played major roles in cleansing the U.S. of its allegedly unfit classes (primarily by sterilization), and they would have cooperated in establishing American race hygiene programs with their German counterparts. It may be true that Davenport's enthusiasm for eugenics was misplaced out of his zeal to make a contribution to society, but it does not exempt him from the damage done to those whose opportunities to marry and have a family were permanently thwarted. Many of them were selected for nonmedical reasons and had the misfortune to be in the wrong social class. I do not consider Davenport and Laughlin to be in the same category of committing evil acts as those major Nazi criminals who faced Nuremberg trials. They never advocated mass murder of the unfit. Their eugenic ideology was muted compared to Nazi ideology.

Notes and References

1. Hannah Arendt, *Eichmann in Jerusalem: A Report on the Banality of Evil* (Viking Press, New York, 1964).
2. For an overall view of the history of allegedly unfit people see Elof Axel Carlson, *The Unfit: A History of a Bad Idea* (Cold Spring Harbor Laboratory Press, Cold Spring Harbor, New York, 2001). For a comprehensive biography and assessment of Laughlin's career see Frances Janet Hassencahl, *Harry H. Laughlin, Eugenics Agent for the House Committee on Immigration and Naturalization, 1921–1931.* Ph.D. thesis dissertation, Case Western Reserve University, Cleveland, Ohio (University Microfilms, Ann Arbor, Michigan, 1969). For a comprehensive and critical obituary of Davenport, see E. Carleton McDowell, "Charles Benedict Davenport: A life of conflicting loyalties." *Bios* **17** (1946): 3–50.
3. Charles Benedict Davenport, *Experimental Morphology*, 2 volumes (Macmillan, New York, 1897).
4. C.B. Davenport, "Dominance in characteristics in poultry," *Report of the Third International Congress of Genetics, 1906* (Royal Horticultural Society, 1907, Rev. W. Wilks, editor; Spottiswoode and Co., London), pp. 138–139.
5. W.E. Castle, "The Reception of Mendelism in America" in *Genetics in the 20th Century*, ed. L.C. Dunn (Macmillan, New York, 1950). See pp. 59–76.
6. Jan Witkowski, *Illuminating Science* (Cold Spring Harbor Laboratory Press, Cold Spring Harbor, New York, 2000). See Witkowski's "Prologue" for a brief history.
7. Ibid. Witkowski has selected 20 articles from the staff between 1903 and 1969.
8. Nicholas Wright Gilham, *A Life of Sir Francis Galton: From African Exploration to the Birth of Eugenics* (Oxford University Press, New York, 2001).
9. Richard Dugdale, *The Jukes: A Study in Crime, Pauperism, Disease, and Heredity* (G.P. Putnam Sons, New York, 1877). Dugdale was born in France and raised in the U.S. His views were Lamarckian, and he saw the Jukes as victims of a bad environment. He died young of rheumatic fever, and after 1880, Weismann's views on the germ plasm replaced Lamarckism (the belief that heredity can be altered in a direct way by the environment).
10. C.B. Davenport, "Degeneration, albinism and inbreeding." *Science* **28** (1908): 454–455.
11. Charles B. Davenport and Morris Steggerda (1929) *Race Crossing in Jamaica* (Reprinted by Negro Universities Press, Westport, Connecticut, 1970). The book first appeared in 1913 as *Heredity of Skin-Color in Negro-White Crosses.* (Most of the field notes were taken by Florence H. Danielson.) Carnegie Institution of Washington Publication 188, 106 pages.
12. H. Nilsson-Ehle, "Kreuzungsuntersuchungen an Häfer und Weizen *Lunds Universitets Arsskrift*" **5** (1909): 1–122; E.M. East, "A Mendelian interpretation of variation that is apparently continuous." *Am. Nat.* **44** (1910): 65–82.
13. C.B. Davenport, *Heredity in Relation to Eugenics* (Henry Holt, New York, 1910).
14. Francis Galton, *Hereditary Genius: An Inquiry into Its Law and Consequences* (Macmillan, London, 1869); Francis Galton, *Natural Inheritance* (Macmillan, London, 1899).

15. D.S. Jordan, "Hereditary Pauperism," in *Footnotes to Evolution: A Series of Popular Addresses on the Evolution of Life* (D. Appleton, New York, 1898).

16. D.S. Jordan, *The Blood of a Nation* (National Unitarian Press, Boston, 1902). D.S. Jordan, *The Heredity of Richard Roe: A Discussion of the Principles of Eugenics* (American Unitarian Press, Boston, 1911). D.S. Jordan, *Unseen Empire: A Study of the Plight of Nations That Cannot Pay Their Debts* (American Unitarian Press, Boston, 1912). D.S. Jordan, *War and the Breed: The Relation of War to the Downfall of Nations* (Beacon Press, Boston, 1915). See also Jordan's autobiography, *Days of a Man: Being Memories of a Naturalist, Teacher, and Minor Prophet of Democracy* (World Book Company, Yonkers-on-the-Hudson, New York, 1922).

17. Alexander Graham Bell, *Upon the formation of a deaf variety of the human race.* National Academy of Sciences. Memoirs, volume 4, number 2, 86 pages. (U.S. Government Printing Office, Washington, D.C., 1884).

18. Davenport, 1910 op. cit., Preface and Introduction.

19. Ibid., p. iii.

20. Ibid., p. iv.

21. Ibid., p. 4.

22. Hassencahl, op. cit., p. 44. Laughlin's papers were donated to the library archives of the Truman State University Library in Missouri. It is a rich collection of correspondence for historians of the American eugenics movement.

23. Ibid., p. 155.

24. Ibid., p. 165. The irony of Laughlin's later life was an onset of epilepsy, once a sign of degeneracy in 19th-century social and medical thinking.

25. Ibid., p. 166.

26. Albert Johnson (1869–1957) was a newspaper publisher in Tacoma and Hoquiam, Washington, who ran for Congress in 1913 opposing citizens' rights for Japanese immigrants and Japanese naturalized American citizens. He eventually chaired the Committee on Immigration and Naturalization until his defeat in 1932. His best-known accomplishments were the 1921 and 1924 restrictive immigration acts, which limited new immigrants to the U.S. for the next 30 years. The 1924 law used the 1890 census (a largely western European majority) for national origins, and this minimized the immigration from eastern and southern Europe.

27. Harry Hamilton Laughlin, *Eugenical Sterilization in the United States.* (Psychopathic Laboratory of the Municipal Court of Chicago, Illinois, 1922).

28. Paul Lombardo, "Three generations, no imbeciles: New light on Buck v. Bell." *NYU Law Rev.* **60** (1985): 31–62.

29. Hassencahl, op. cit., p. 339.

30. Ibid., p. 329.

31. Madison Grant (1865–1932) was a lawyer in New York City and a friend of Davenport. He took an interest in the conservation movement and helped establish the Bronx Zoo. He was virulently racist and anti-Semitic, and his book, *The Passing of the Great Race, or, the Racial Basis of European History* (Scribner's, New York, 1916), was a best-seller and translated into several languages, including German. It was much admired by Hitler, who called it "my bible."

32. Hassencahl, op. cit., p. 344.

5

Heroes with Feet of Clay
Francis Galton and Harry Clay Sharp

DAVENPORT WAS OPINIONATED, MILDLY BIGOTED, uncritical of the evidence he used to bolster his distaste for those with "bad protoplasm," and power-hungry for influence among the rich and famous. Those are combinations that might have made him dangerous had the U.S. moved in the same direction of racial hygiene as had Germany. Lots of things prevented this. Foremost was the heterogeneity of the population of the U.S. Germany was far more homogeneous (less than 2% Jewish) and thus subject to a racial flattery and self-image that the U.S., an immigrant country, lacked. Despite my bleak assessment of Charles Davenport's applied genetic outlook, I think he is less likely to have emulated Adolph Eichmann as an ardent bureaucrat for an American Holocaust than Harry Laughlin. Davenport felt ambivalence to compulsory sterilization. His need for approval was also a major facet of his personality, and if his peers expressed hostility to his ideas he would have tried to compromise (as he did in muzzling Laughlin) or fall silent unless the criticisms were savage. I think Davenport would have delegated that work to Laughlin and would have defined his role like that of Eugen Fischer or Fritz Lenz—a theoretician who did not do the dirty work. Laughlin had fewer credentials as a scientist than Davenport. Like Eichmann, Laughlin started out as an insignificant person and realized he could become powerful and famous only if he had the opportunity to implement the beliefs of those whose cause he had joined. His beliefs were steeped in bias—he was anti-Semitic, took a dim view of eastern and southern Europeans, and did not for a moment flinch at the thought of sterilizing at least 10% of the U.S. every generation, weeding his way through the bottom muck of the "unfit" until he and his hoped-for followers would have purged it of this alleged blight of bad genes. He attached himself to Davenport because he admired his talent and his power.

Quite different are those whose bad outcomes are associated with more successful careers or with motivations that are less steeped in bigotry than they are in the reformer's zeal. Both Francis Galton and Harry Clay Sharp differ from Hannah Arendt's accusation of practicing the banality of evil. Galton was almost a religious reformer in his idealism for the progress of civilization. Sharp was a crusader for public health. Each had an opportunity for a successful career without eugenics. Laughlin had no such talent, and Davenport felt committed to his empire, and one of its strengths, he believed, was the eugenics movement.

Galton's Multiple Interests as a Generalist

Galton came from a wealthy family. He was also connected to the Darwins and Wedgwoods, Charles Darwin being his first cousin.[1] His father was a Quaker and a bit of a crank, who made his fortune selling guns. He justified his occupation and was forced to quit the Quakers, whose pacifist views were at odds with his awkward dualism. As a young child, Galton was tutored by his older sister, who was forced into a sedentary life because she was crippled. She made him into a child prodigy, and he doted on her attention and love. He came to believe he was a special gift for the world and proudly achieved whatever tasks she assigned to him. When he went to college, he hoped to major in mathematics, but he suddenly discovered he was only of limited talent in that pursuit and he could not compete for the top honors. It was a bitter defeat for his ego and he felt embarrassed that he had let his family down. He switched to medicine. Shortly before his graduation, his father died and left him an immense fortune. Galton decided he would use the money to subsidize himself as a scholar and that he had no need to practice medicine after his graduation.

Galton's first venture was to provide funding for an expedition to Africa. This was the age when British colonies were forming rapidly around the world, and there was a strong demand to learn about the cultures, geography, and resources of relatively unknown areas of the world. He wrote his first book, on traveling, which turned out to be a useful manual for explorers.[2] His studies of the geographical features in sub-Saharan Africa were well received, and he got himself a gold medal from the Explorer's Club. Galton turned his attention to meteorology and described the anti-cyclone and how Coriolis forces lead to mirror-image weather patterns below the Equator. He took an interest in statistics and began to apply mathematics to almost all observable phenomena. He invented the correlation coefficient and showed how this could be used to identify potential causal relations.

Galton never made a major discovery of the same scale as his cousin Charles, but he was recognized as a solid scientist whose publications were followed with interest. Unlike his cousin, he chose to be a generalist and to explore many different fields rather than commit himself to one. He introduced fingerprinting for identification and worked out a method to classify them (loops, whorls, ridge counts, creases, angles of intersecting radii, etc.).[3] He even demonstrated, using childhood palm prints kept by parents, that the fingerprint patterns did not change with age. He had some original ways of doing things. He would make composite or average faces by partial exposures of faces set to the same size and showed how these approached the idealized view of beauty. He found no particular facial features characteristic of criminals. In this attempt, he was guided by the widespread belief in physiognomy and other attempts to classify human behavior by anatomic differences, especially facial features and skull shape.

Sometimes Galton's studies were motivated by a perverse sense of humor or deep skepticism about social beliefs. He wanted to see if he could conjure up a religious feeling of reverence by selecting an unlikely idol.[4] He chose the character, Punch (also called Chiarivari), and bought a photo of the clown in a tricorn hat and placed it on the mantel in his bedroom. Each night he would pray to it and revere it. Within a month or two, he became quite attached to this idol and felt anyone could revere anything just by trying to do so. In fact, he remarked that much later in life if he happened to pass a shop with a Punch doll in the window, a feeling of reverential awe would fill his body. Galton's curiosity about human peculiarities is discussed in his autobiography. He never published most of these experiments that revealed trivial facets of human faculties. He did publish an article on the efficacy of prayer and claimed that public prayer, offered for the royal family every Sunday throughout the world's Anglican churches, had little effect on extending the mean life expectancy of the royal family. He noted that "the sovereigns are literally the shortest lived of all who have the advantage of affluence." He concluded that public prayer was ineffective either because it was insincere or because God had other things to do rather than respond to public prayer.[5]

Galton's eugenics views were tied to his interest in heredity. Because he was a prodigy of sorts, and certainly had a wide range of interests in which he held his own, he felt he owed his gifts to his family history. There were a lot of famous people in his family. He turned his attention to two skills—competitive athletics (such as sculling on the Thames) at Oxford and Cambridge, and the high honors, especially in mathematics of the final year contestants. He showed that there were a considerable number of first-degree relatives who were among these champion performers. He then studied genius in the British population by two approaches. He measured the length of obituary columns for 1 year in the *London Times* to obtain the incidence of those that he classified as illustrious, eminent, well-known, locally known, or those with ordinary careers but no special unique contributions. To classify these categories, he also used Chambers' *Dictionary of Biography*. He concluded that eminent people appear once among 4000 people in the population. Illustrious persons were about 1 in one million (persons for whom the whole world of intellect weeps, upon their death, in his description). Galton published two books on the inheritance of genius and claimed it was transmitted to about 20% of primary relatives.[6] His control for this study was curious. He used the nepotically raised sons placed in the households of bishops and cardinals by friends or relatives of the prelate. Death in childbirth was not rare in those days, and this gave the sons an opportunity to work as household staff and benefit from the music, conversations, and stimulating people who came to visit. The virtual absence of such adopted children appearing in either the obituary notices or the dictionaries of biography suggested to Galton that heredity, and not an enriched environment, was the key factor in the lives of the eminent.

In 1883, Galton coined the term "eugenics."[7] He believed it was a moral duty for those who are talented to have a larger number of children than the average family. He saw eugenics as a secular religion that would be motivated by the good that people of talent bring to the world. Galton accepted a view of history that people of heroic stature shape history. When we think of the 19th century, we name Lincoln, Napoleon, Bismarck, Darwin, Beethoven, Van Gogh, Pasteur, Mendel, Faraday, Dalton, Mill, Tolstoy, and Lister as among the great contributors to its legacy. If these illustrious people and scores more of eminent ones contributed so much to civilization, why not encourage people like them to have many more children and raise the number of such future geniuses in society? With a transmission rate of 20%, that is a highly achievable result, compared to the 1 in 4000 or 1 in a million for such persons to arise randomly.

Today, we call Galton's eugenics "positive eugenics" because it favors the best individuals breeding with those similarly ranked as the best. In his last years of life, the eugenics movement, especially in the U.S., was stressing negative eugenics—the elimination of the unfit. The U.S., Davenport and Laughlin included, rarely stressed going for the rare geniuses, and the American eugenics movement was more comfortable appealing to people who were just a notch above mediocrity, as those winning "fitter family" contests at 4H fairs would demonstrate.[8] Unfortunately, in the early 20th century, Galton was old and vain and endorsed these efforts to contain the unfit through negative eugenics measures. He had tried, without success, for some 30 years to encourage his peers to breed, but although they applauded his ideals, they felt uncomfortable giving up their time and luxuries to raise large families.

Galton does not rate as an illustrious person, although he certainly was eminent. He was a reasonably objective scientist. One of his sad triumphs was doing the experiments that disproved the theory of heredity championed by his cousin, Charles Darwin.[9] He carried out the transfusions of pigmented and nonpigmented rabbits to see whether "gemmules," the alleged bits of heredity given off by somatic cells and ending up in reproductive cells, did contaminate the heredity of these strains. He found no such transmission through the blood, and although Darwin tried to rescue his theory, for most biologists who read Galton's papers, Darwin's provisional theory of pangenesis with its gemmules was dead.

One cannot dismiss Galton as banal, like Davenport and Laughlin. He was a better scientist than they were, and he had a consistent integrity for much of his work. How then does one classify this father of eugenics? He did not invent negative eugenics, and his later endorsement was almost an afterthought of an old man who was out of it. In his more vigorous years, his ideas of eugenics were not to harm people but to encourage the brightest and healthiest to have more children. It would be unfair to judge him by current standards of knowledge about the inheritance of behavioral traits, although we can certainly question his judgment in allowing himself to comment on eugenic policy at a time when he lacked his full capacities. One can, however, criticize him as flawed, because he used a shoddy control for his evidence for a 20% transmissibility of genius and because he was not critical of the objectives, criteria, or plans of implementation of the negative eugenics movement that arose later. For these reasons, I consider Galton less malevolent in the bad outcomes of eugenics than Davenport or Laughlin. He is a hero with feet of clay.

Harry Sharp Puts Medical Idealism to Practice

Harry Clay Sharp was born and raised in Indiana, but he attended medical school in Louisville, Kentucky. He received his M.D. in 1894 and took a job as prison physician across the Ohio River in nearby Jeffersonville, Indiana. A reformatory and prison had been established there shortly after statehood in 1818, and prisoners were used to build the canals along the Ohio River. When Dr. Sharp looked over the prison, he was appalled. There was no proper kitchen, and the prisoners were malnourished and ate what amounted to slops. There were no private privies in the cells, and prisoners would toss their body wastes from a pot into a trench that ran down the corridor between the banks of cells. Each morning a guard would hose down the wastes and flush them outdoors. Sharp helped establish an annual report from the prison to the governor. It was printed by a press at the prison. These reports also went to state supervisory offices.[10] Sharp criticized the state for neglecting its prisoners. He believed the deaths from infectious diseases such as typhoid were preventable and would not occur if there were privies in the cells. He felt his prisoners could survive many other infections if they had a more balanced diet. He pointed out that if this had happened in an institution other than a prison, the governor and other officials could be charged with negligent homicide.

Dr. Sharp got his requests, and he happily made his tables reporting the surgeries, medical treatments, and diseases of his prisoners and praised the governor for his prompt action. In 1899, Dr. Sharp read an article in the *Journal of the American Medical Association* by a prominent surgeon in Chicago, Albert Ochsner.[11] Ochsner described a surgical procedure, vasectomy, that he performed on two middle-aged men with enlarged prostates. Swedish and English surgeons in the 1890s believed (erroneously) that vasectomies reduced the size of prostate glands. Ochsner noted that both of his patients resumed an active sex life after surgery, and, other than sterilization and presumably smaller prostates, he found no additional effects from the surgery. He concluded his paper with the remark that if vasectomies were performed on degenerates, the incidence of unfit people would dramatically be reduced.

Dr. Sharp had learned in medical school that masturbation (then called onanism) led to physical and mental degeneracy as well as debilitated offspring.[12] He believed he should caution his prisoners about masturbating. One of his prisoners asked to be castrated because he could not stop masturbating and feared he would die. Dr. Sharp performed his first vasectomy as a treatment for masturbation and claimed that it benefited his patient, who gained weight and took on a more optimistic attitude toward life. He began reporting more cases of vasectomy as medical treatments for onanism in his annual reports. He also wrote letters to the governor asking him to introduce a bill to permit state sterilization of degenerates. In 1909, both houses of the legislature and the governor's signature brought the first legal sterilization act into being. It is not clear how many men Dr. Sharp sterilized. He claims it was about 600, but some of his figures might be inflated; they are certainly inconsistent based on the annual reports. Dr. Sharp also presented his findings in a number of published papers, some given at the American Medical Association meetings in Atlantic City, New Jersey.[13] He began a postcard campaign urging his fellow physicians in the U.S. to petition their governors and representatives for compulsory sterilization laws.

Dr. Sharp's good intentions are mixed. No one would fault his concern about the health of his prisoners. In the case of the typhoid fever infections (which ended after privies were introduced into the cells), he would be applauded as a caring physician who took on an indifferent governmental attitude and brought about necessary reforms. Nor would one who is familiar with medical history consider his views about onanism the inspired work of a madman. Text books written by Bellevue physicians and used in medical schools warned physicians (as they had been warned since the mid 1700s) that masturbation was a major cause of mental and physical illness. One could fault him for not doing controlled experiments. One could fault him for urging the state to do this against the wishes of the prisoners. Dr. Sharp was raised in a tradition of medical paternalism that persisted to the 1970s. In that tradition, physicians knew best what was good for their patients, and their patients had no rights to interfere with that oath-bound medical judgment for the patient's welfare.

Dr. Sharp's enthusiasm with his colleagues worked. About 30 states enacted compulsory sterilization laws. Some were struck down by state courts, and eventually Laughlin helped design a model eugenic sterilization law that was tested in Virginia and engineered to move its way to the Supreme Court in 1927.[14] The court rendered an 8–1 verdict supporting the right of the state of Virginia to sterilize a patient, Carrie Buck, portrayed as a third-generation imbecile. Both Carrie and her daughter were sterilized. By 1940, more than 40,000 men and women had been sterilized against their will as degenerates, most of them psychotic or mentally retarded. That is a significant bad outcome for Dr. Sharp's good intentions.

In 1937, Dr. Sharp was interviewed for an article on eugenic sterilization in the *Journal of Heredity*.[15] It was also a year anticipating the preparation for yet another war with Germany with a considerable bad press for the excesses of Nazi Germany. The Nazis had used the model eugenic law in drafting their own sterilization laws. Nazi race hygiene offended most of the civilized world, and Nazi Germany was applying its eugenic laws on a massive scale. Dr. Sharp had long moved away from Indiana, and, after serving in the army as a surgeon in World War I, he returned to medical practice and found a position in an asylum for the retarded in Vineland, New Jersey. He was well aware of the feelings against Hitler's Germany and was himself appalled by Nazi ideology. When asked how he felt about being the father of compulsory sterilization laws, he replied "We didn't know enough science then."[16]

I have reflected on Dr. Sharp's comments many times, wondering what he should have replied to that question. His answer suggests he had not examined his values, and that, in principle, sterilizing the unfit was morally consistent with his medical oath to do no harm. By blaming science (e.g., Ochsner's paper) instead of his own values, he becomes the hapless soldier on trial for following orders and killing civilians in My Lai during the Vietnam war. He becomes the pharmaceutical scientist who cooks up a new test to make a product look better than its initial test results which indicated it didn't work. He obviously felt guilty for the bad outcome and was ending his career not in glory, but in a cloud. He was certainly not a Mengele. He believed in treating, not killing, his patients. He also must have felt confused, because at the time when he coupled vasectomy and degeneracy through onanism, he was making a virtually untested theory serve as a vehicle for state action. At the same time, he saw himself as a person who was trying to do good for his patients, and by urging government action in the late 1890s he had helped to save lives through an improved kitchen and through individual privies in prisons. Acting with unexamined or poorly examined science is a moral lapse. Acting in ignorance or error is a personal calamity but not necessarily a moral lapse. By blaming science instead of himself for his hasty implementation of an untested medical procedure (using vasectomies as a treatment for prevention of masturbation), Harry Sharp revealed his feet of clay.

Notes and References

1. Nicholas Wright Gilham, *A Life of Sir Francis Galton: From African Exploration to the Start of Eugenics* (Oxford University Press, New York, 2001).

2. Francis Galton, *The Art of Travel; or Shifts and Contrivances Available in Wild Countries* (Murray, London, 1855).

3. Francis Galton, *Finger Prints* (Macmillan, London, 1893).

4. Francis Galton, *Memories of My Life* (Methuen and Company, London, 1908), p. 277.

5. Ibid. Among other curious experiments cited in his autobiography that Galton did are his measurement of boredom at public lectures by lining up the audience with the side of an index card and counting the number of nodding heads in the slice of rows his card cut through (p. 278); his experiment to see whether besides sound ("two and two are four") and vision ("$2 + 2 = 4$"), other senses could be trained to do mathematical operations. He succeeded with smell, using different scents to represent numbers, and within a week he was adding and subtracting by scent (p. 284); and his use of a dog whistle secreted in a hollowed-out cane that he would activate while at a zoo and noted how it disturbed the lions in their cages (p. 247). He does not mention in his autobiography his work on public prayer. Perhaps he was embarrassed by the negative response to it. Francis Galton. "Statistical inquiries into the efficacy of prayer." *Fortnightly Review* **68** (1872): 125–135 (see p. 127).

6. Francis Galton, *Hereditary Genius: An Inquiry into Laws and Consequences.* (Macmillan, London, 1869); *Natural Inheritance* (Macmillan, London, 1889).

7. Francis Galton, *Inquiries into Human Faculty and Its Development* (Macmillan, London, 1883).

8. Steven Selden, *Inheriting Shame: The Story of Eugenics and Racism in America* (Teachers College Press, Columbia University, New York, 1999).

9. Francis Galton, "Experiments in pangenesis, by breeding from rabbits of a pure variety into whose circulation blood taken from other varieties had previously been largely transfused." *Proc. R. Soc.* **19** (1871): 393–404.

10. *First Biennial Report of the Board of Managers of the Indiana Reformatory*, Jeffersonville, Indiana from November 1, 1896 to October 31, 1898, inclusive.

11. Albert John Ochsner, "Surgical treatment of habitual criminals." *Am. Med. Assoc.* **32** (1899): 867–868.

12. Elof Axel Carlson, *The Unfit: A History of a Bad Idea* (Cold Spring Harbor Laboratory Press, Cold Spring Harbor, New York, 2001). See Chapter 3, Self-pollution and declining health.

13. Harry C. Sharp, "Vasectomy as a means of preventing procreation in defectives." *Am. Med. Assoc.* **53** (1909): 1897–1902.

14. Paul Lombardo, "Three generations, no imbeciles: New light on Buck v Bell." *N. Y. Univ. Law Review* **60** (1985): 31–62.

15. William M. Kantor, "Beginnings of sterilization in America." *Heredity* **28** (1937): 374–376.

16. Ibid., p. 374.

PART 3

WAR TIME AND THE THREAT OF WAR AS JUSTIFICATION FOR SUSPENDING ETHICAL AND MORAL BEHAVIOR BY SCIENTISTS

D<small>URING A WAR OR THE THREAT OF WAR,</small> scientists may find themselves applying basic science to the design of military weapons. The recruitment of such scientists is easy because patriotism is part of the culture of every country, and many citizens, whatever country they live in, believe in a value represented by the excerpted phrase "my country, right or wrong." For this reason, American scientists willingly moved to Los Alamos in New Mexico to work on the top-secret development of atomic weapons codenamed the Manhattan Project. Their German counterparts were also working on the design of nuclear weapons.

American physicists at Los Alamos in New Mexico were convinced that Germany was working on nuclear weapons and that the Nazis would use those weapons to force an end to the war. They were divided on whether to use such weapons after the defeat of Germany in May, 1945, because they doubted that the Japanese had the technology to develop their own nuclear weapons. I explore those moral issues that arose and the effect it had on the careers of those involved in nuclear weapons production.

I also explore the use of the herbicides 2,4-dichlorophenoxyacetic acid (2,4-D) and 2,4,5-trichlorophenoxyacetic acid (2,4,5-T), which were used extensively in the Vietnam war. Unlike the development of nuclear fission, which immediately was shifted from basic science in 1938 to wartime applications after 1939, the development of 2,4-D and 2,4,5-T in World War II came too late for extensive military use; a civilian use prevailed until the Vietnamese war in the 1960s employed massive amounts of these agents in a mixture called Agent Orange. Unlike Los Alamos, where scientists worked with the military, Agent Orange production and use was almost exclusively a military operation with limited input from ecologists or those concerned with the hazards of chemical agents to civilian or military health.

In both cases, it is the state, under the conditions or rationale of wartime necessity, that determines the outcome of applied science, but in both cases, a sizable workforce of technicians and scientists is involved to produce weapons in large quantities.

6

Radiation in Peace and War

THE RADIATION CONTROVERSY CAN BEST be explored by dividing it into three historical phases. The first covers 1895–1940, when early medical and biological effects of ionizing radiation were first noted and described. The second phase covers the war years, 1939–1945, when Germany, Japan, Great Britain, and the U.S. established programs to develop atomic weapons. The third phase covers the Cold War years, 1946–1990, when a nuclear arms race dominated the values and expectations of the world's major countries and nuclear reactors became a major source of supplying electric power.

To interpret these three phases, several factual features should be kept in mind:

1. X rays passing through air or water produce ions by stripping electrons off atoms. They also impart energy to atoms. This makes the atoms chemically reactive.

2. X rays passing through living tissue cause two types of genetic damage. One type involves breaking chromosomes. The other type involves a chemical or physical alteration of a gene.

3. In general, chromosome breaks in sperm or eggs lead to infertility or aborted embryos or malformed births. In adult body tissues, they may lead to mild or severe radiation sickness; later in life, they may lead to cancers.

4. Gene mutations, in general, are not expressed in the children of an exposed parent. They show up several generations later in a population. That is because most induced mutations in an egg or sperm are recessive and require both parents to introduce the same mutant gene

into a child, a situation normally found among descendant cousins or inbred populations.

5. Gene mutations are cumulative and proportional in frequency to the ionizing radiation dose we receive.

6. A chest X ray requires about 0.1–0.01 roentgens. This is a low dose. A Hiroshima or Nagasaki survivor of severe radiation sickness had a whole-body dose of 100–300 roentgens. The mean-lethal dose in Hiroshima (a whole-body exposure in which half of the exposed human population dies) is about 450 roentgens. These are high doses.

7. Physicists and geneticists agree on the hazards of high doses of ionizing radiation. They differ on biological effects of low doses.

Phase I: The Discovery and First Uses of Ionizing Radiation

Wilhelm Roentgen discovered X rays in 1895 and was amazed that he could see objects in closed boxes. Within a year after the announcement of his discovery, X rays were being used in hospitals to locate foreign objects in bodies, to identify fractures in bones, and to diagnose tumors. As physicists and health professionals found applications to their work, reports began to appear of unexpected consequences among these early X-ray pioneers. They reported reddening of the skin, blistering, and ulceration that seemed to persist, healing very slowly. By the early 1900s, many X-ray users complained of permanent numbness and damage to exposed fingers and hands and the appearance of tumors. Dozens of these early workers died of the cancers they had induced. Radiation workers up to the 1920s used a physiological approach to safety, defining a "tolerance dose" as one that fell below reddening of the skin. It would not be until the late 1920s that effective dosimeters were used to measure, quantitatively, the doses applied to patients and the doses received by practitioners.

In these early years of 1896–1910, biologists noted X rays applied to eggs or sperm of amphibians or rats at high doses caused embryonic damage, aborting the fertilized eggs. At very high doses, the X-rayed sperm produced parthenogenetic development of fertilized toad eggs. At lower doses the embryos were normal or produced less drastic damage to the tadpoles or adult amphibians that emerged.[1]

Geneticist Herman Joseph Muller, a member of Thomas Hunt Morgan's fruit fly genetics laboratory, received his Ph.D. in 1915. He chose as his life's work the related themes of mutation and the gene. While at the University of Texas in Austin, in 1926, he began a series of high-dose radiation exposure of male fruit flies; he designed a genetic stock for these experiments that would recognize a category of mutations that cannot be seen directly because they killed the embryos expressing them. These were called X-linked recessive lethal mutations. As the name implies, a mutation on the male X chromosome in a sperm will fertilize an egg with genetic markers on her X chromosome (the mutation *bar eyes*, a dominantly expressed mutation, a recessive lethal that arose independently in 1921, and an inverted segment of the X chromosome containing these markers). The marked X chromosome was called ClB (pronounced "see ell bee"). Each daughter containing the ClB chromosome and the X-ray-exposed X chromosome from the sperm was mated with males with another set of visible marker genes free of the inversion. The resulting offspring from each mating in a vial should give two classes of sons, those with the ClB (necessarily aborted because they contain the 1921 lethal) and those with the X-rayed X chromosome. However, if the X-rayed chromosome in that sperm had a newly induced X-linked recessive lethal mutation, that category of males would also be absent and only females would be present among their progeny in the vial. The females with the appropriate induced lethal could then be used to map the location of that induced mutation. In some cases, Muller got vials containing induced X-linked visible mutations among the males, and these too could be mapped. Muller estimated that the induced recessive lethals were 10 times more frequent than the induced recessive visibles.[2]

In 1927–1931, when Muller presented his results in a series of papers, he realized that he had founded a new field of radiation genetics. In addition to showing a linear response of induced mutations to X-ray dose, Muller and those who joined in exploring this new technique found another event associated with X rays—they induced chromosome rearrangements (called inversions, translocations, duplications, and deletions). These seemed to rise in frequency exponentially as the dose increased. A third phenomenon noted at higher doses was a large number of "dominant lethals," fertilized eggs whose embryos never developed. The cause of these remained unknown until 1940 when Muller and his student, Guido Pontecorvo, demonstrated that chromosomes with two centromeres (dicentric chromosomes) induced by X rays were aborting the embryos.[3] Independently, Barbara McClintock, using maize, had found a related effect of such dicentric chromosomes and showed that they arose from an initial spontaneous break of a chromosome that was not repaired until after the chromosome had replicated. She called this a breakage-fusion-bridge cycle.[4] In his 1940 papers, Muller referred to this mechanism as the cause of "radiation necrosis," the term applied in the early 1900s by physicians describing the harmful effects of radiation on normal human tissue.[5] But radiation necrosis also was beneficial when that damage occurred in X-rayed tumors. Radiation was thus perceived as having beneficial uses (diagnoses and treatment of tumors); detrimental effects (radiation necrosis of healthy tissue, induced mutations in reproductive cells, induced rearrangements in reproductive cells, and the induction of cancers); and questionable applications (the use of X rays to temporarily sterilize males or to induce ovulation in infertile females, the treatment of plantar warts, straightening out of bowed legs in children, and measuring foot size by shoe-store personnel to get a best fit for customers).

One other finding, in 1939, was very important for Muller's interpretation of radiation effects when applied to humans. With his student S.P. Ray-Chaudhuri, in Edinburgh, he found that it made no difference whether a moderate or high dose of radiation (500 or 2000 roentgens) was applied to fruit fly sperm for a short duration of time (30 minutes) or over a long period of time (1 month).[6] The total number of X-linked lethal mutations induced for the two populations was the same at a given dose. This suggested to Muller that so-called tolerance doses were not valid and that medical practitioners had to protect themselves and their patients from unnecessary exposure to low doses of radiation. At the 30-minute applied dose for 500 roentgens, the rate is about 17 roentgens per minute; for the 30-day applied dose, the rate is 0.01 roentgens per minute. Muller's first warning on radiation hazards, given in Waco, Texas in 1928, was poorly received by some practitioners, who walked out in anger. An equally negative response came from practitioners in Great Britain, who criticized Muller for alarming patients who might avoid effective diagnosis or treatment if they looked upon radiation as hazardous to their health.

After the atomic bomb was used at Hiroshima and Nagasaki, Muller interpreted the reports of radiation sickness as an example of induced radiation necrosis caused by breakage-fusion-bridge cycles in the dividing cells of adult tissues (skin, bone marrow, and the lining tissues in the lumens of the digestive tract and the circulatory system). He was also given his Nobel Prize in 1946, and this made him a spokesman for radiation protection in the new Atomic Age.

Phase II: The Making of the Atomic Bomb

Physicists in the 1930s knew each other and followed each other's work with interest. Of particular interest were the study of atomic structure and the instability of atoms when exposed to the products of radioactive decay. Some particles, including protons or neutrons, emanate from naturally occurring elements, such as radium. In 1938, Lise Meitner, working in Otto Hahn's laboratory, discovered she could split atoms with neutrons, converting heavier atoms into lighter ones, along with the release of additional neutrons and energy in the form of gamma rays. This paper suggested to many physicists, including Albert Einstein and Leo Szilard, that an attempt might be made to create an explosive weapon—an atomic bomb—if fissionable material (especially uranium isotopes) could be amassed. Szilard urged Einstein to write President Franklin D. Roosevelt that Nazi Germany was very likely to initiate such a program and that the U.S. might wish to begin a program of its own at least as a deterrent.

The Manhattan Project was started shortly after Roosevelt consulted with his military advisors. A site was selected at a boys' school on the high, isolated mesas of Los Alamos in New Mexico to house a secret facility. This was a remote and well-guarded region for beginning the design and construction of atomic weapons. The major scientific figures in the organization of the Los Alamos laboratory included J. Robert Oppenheimer, Ernst O. Lawrence, Kenneth Bainbridge, and Edward Teller. A "who's who" of prominent scientists were recruited, and they included Philip Morison, Robert Wilson, Hans Bethe, Willard Libby, and Richard Feynman. Oppenheimer was the scientific director for the program, and the entire project was supervised by General Leslie Groves, who established security policies and the overall engineering coordination for the components of the bombs, which was extensive.[7] This included procurement of devices to concentrate and isolate the very fissionable uranium-235 isotope and plutonium from the Hanover facility in Washington state and the Oak Ridge facility in Tennessee. All of the top scientists at Los Alamos knew what they were doing by 1943, when the major features of putting a weapon together had been worked out. Discussions about the implications of using such a weapon began to circulate as the reality of the project emerged, and what might have originally been a remote possibility had become a likely achievement.

The range of views was quite wide. Bainbridge (Harvard University) was in charge of the Trinity project, which was the first successful device to show the explosive power of an atomic weapon. It was exploded near Alamogordo, New Mexico, in July, 1945. Bainbridge favored using the bomb as a demonstration on a model town built in an isolated area and inviting the Japanese military to observe it.[8] The U.S. military's argument against this was that it would take too much time and, in the meantime, U.S. casualties would continue to be high from the fierce resistance the Japanese displayed in their southern islands. Lawrence (University of California-Berkeley) argued that "the bomb will never be dropped on people. As soon as we get it, we'll use it only to dictate terms of peace."[9] Oppenheimer knew of moral doubts from many of his fellow physicists at Los Alamos, but he withheld those feelings when the President's special committee convened (consisting of Secretary of State James Byrnes, Vannevar Bush, James Conant, Secretary of War Stimson, and General George Marshall). Oppenheimer did not believe the bomb would kill more than 20,000 people; he had considerably underestimated the bomb's destructive power. He also believed that the casualties would not be worse than fire bombings already used on Tokyo and other large industrial cities in Japan.[10] In this respect, his views were similar to those of Stimson, who thought there was no difference between atomic bombs and conventional bombing that had been used in the war.[11] Marshall wanted to hit a naval base after giving advance notice to clear the area of civilians and military personnel. He felt this would give the U.S. "the moral value of giving the Japanese advance notice."[12] A major reason that demonstrations of various kinds were rejected was fear of a dud, which would embolden the Japanese and which would, if it were picked up by the Japanese, be modified and used by them. Byrnes gave the information to President Harry S. Truman, who agreed with the military assessment that a demonstration was impractical. Truman thought the bomb should be used to shorten the war.

Szilard, who had initiated the American project to build a bomb against Hitler, felt less compelled to use it against the Japanese. He tried to get a petition signed by physicists involved in the project but couldn't get many to sign. Oppenheimer blocked the presentation of that petition by giving it to Groves, who filed it away with the understanding that it would not be read by Truman. Although some scientists were troubled and did not want the bomb to be used, others sought rationalizations to justify their approval of using the bomb. Teller took a different view: "I have no hope of clearing my conscience. The things we are working on are so terrible that no amount of protestation or fiddling with politics will save our souls."[13] A similar view was offered by Bainbridge as he watched the device explode at Alamogordo and told Oppenheimer, "Oppie, now we're all sons of bitches." Oppenheimer was overheard saying something different by *New York Times* science reporter William Laurence: Oppenheimer quoted from the *Bhagavad Gita*, "Now I am become death, the destroyer of worlds."[14]

As the previous paragraphs demonstrate, rationalizing one's wrongdoing is more frequently attempted than acknowledging that what one is doing is wrong. Compton used the argument that the atomic bomb was necessary for a "probable net savings of lives."[15] Oppenheimer shared Groves's view that the use of the bomb would "drive home its full impact." They believed there was "no acceptable alternative to direct military use."[16] Lawrence naively hoped, "Now we will have no more war and the most backward countries will be able to start catching up."[17] Ironically, one month after the bombs were dropped on Hiroshima and Nagasaki, the U.S. Army Air Force drew up a list of 15 Soviet cities as targets if a future war with the USSR emerged.[18] As the reality of his work sank in and was taken out of his hands, Oppenhiemer's mood changed and he expressed second thoughts, "If atomic bombs are to be added as new weapons to the arsenals of a warring world, or to the arsenals of nations preparing for war, then the time will come when mankind will curse the names of Los Alamos and Hiroshima."[19]

Oppenheimer's change of heart, as well as his opposition to the development of the hydrogen bomb, led to his ousting in 1954 as a security risk from the Los Alamos program. Although Groves was aware of Oppenheimer's left-wing views in the late 1930s before the Hitler–Stalin nonaggression pact, these were not held against him during the war. During the Cold War, especially in the 1950s, Oppenheimer was perceived as a former Communist not to be trusted. His family included known members of the Communist Party. When I interviewed Willard Libby in 1971, he told me that Oppenheimer tried to get him and other scientists to come to Communist-affiliated groups on campus (see footnote 21).

The arguments regarding the use of nuclear weapons can be gathered into the following groups:

1. *Don't use the bomb because it is too powerful and the major casualties will be innocent civilians. Killing civilians is wrong.* We can call this the moral argument against killing (and it includes both religious commandments not to kill and Kantian ethics).

2. *Use the bomb even if it kills a substantial number of innocent civilians because this will shorten the war and save more lives than would be lost*

by conventional warfare. We can call this a utilitarian argument involving the greatest good for the greatest number.

3. *Use the bomb because it will scare the wits out of future leaders who will recoil from using weapons of mass destruction.* I will call this argument the belief that national leaders give in to or shun overwhelming force or terror.

4. *Use the bomb because our enemies attacked us savagely and do not deserve any sympathy from us.* This is the principle of revenge or justice.

The debate over the use of the atomic bombs in 1945 continues with adherents parceled out among these different moral positions. During the war, the U.S. and its allies did not know how far along Japan and Germany were in their attempts to design and build atomic weapons. The U.S. and Great Britain cooperated between the Los Alamos and Harwell facilities. A major figure in that cooperation was Klaus Fuchs, who would later be revealed as an agent for the USSR and who passed U.S. weapons information to them. Fuchs was a major contributor to the design of the bomb himself. After the war ended, it was clear from secretly taped conversations of the imprisoned German physicist Werner Heisenberg and his colleagues that the German effort had failed despite good-faith efforts to make it work. Japanese physicists working on the project were even farther behind than the Germans. From the political perspective, Truman's decision to use the bombs on Hiroshima and Nagasaki has been analyzed numerous times. Some claim he felt obliged to use the weapons to keep the USSR from staking a major claim in the Far East if they joined the planned Allied invasion of Japan. Some feel he was concerned that the planned November, 1945, invasion of mainland Japan would cost up to 300,000 American lives. In all likelihood, it was the concern that a blockade alone would not work and that Japan would have to be invaded at a heavy cost in lives that tipped Truman's decision.

One final issue has rarely been a part of the debate about the use of the atomic bombs in World War II. How much did the military, the President's advisors, and the physicists at Los Alamos know about the biological effects of radiation? Muller told me when I was his graduate student that he was a consultant for the Manhattan Project and gave advice about the biological effects of radiation. None of that information was in the Muller archives at the Lilly Library at Indiana University (although some of the archival material may have been put in a delayed release category, or he might have returned such items, classified secret, to the Manhattan Project according to their instructions). It is clear from the numerous books on the making of the bombs that biological effects were not an issue in the moral debates about the uses of the bomb. The major concern was blast effects killing civilians.

Phase III: The Cold War and the Shaping of Moral Argument

After Muller received his Nobel Prize in 1946 for founding the field of radiation genetics, he was in heavy demand as a speaker on the hazards of radiation. He was also a member of numerous national and international committees on radiation safety. There were three concerns in Muller's approach to the radiation controversy. Of first importance was the proliferation of nuclear weapons, an event virtually all scientists knowledgeable about how science works knew would emerge with other nations seeking to develop the prestige and protection of being a nuclear power. Second was the concern over the medical use of radiation and the indifference by practitioners to their own safety and to that of their patients. Low doses of radiation were not considered harmful. Third was the potential industrial use of radiation, including plans by physicists to harness the "peaceful atom" and supply cheap, unlimited amounts of electric power using nuclear reactors.[20]

Muller ran into difficulty on all three approaches. He was considered suspect by the security agencies involved in protecting the supremacy of an American and British monopoly on nuclear weapons. Muller had been a Communist in all but membership from the 1920s through 1936, when he fled the USSR after his debates with Lysenko. His anti-Soviet views after 1948 were looked upon in the 1950s as a ruse to carry out a secret mission to sabotage U.S. weapons development by raising concerns about radiation damage.[21] He was also considered an extremist by those health practitioners who saw no harm from low doses of radiation and who felt any concerns about radiation safety by patients would lead them to neglect essential diagnostic and therapeutic treatment. Muller was also looked upon as an obstructionist when he insisted that nuclear reactors start out with stringent standards of low doses of emissions, because once the reactors were built, it would be impossible to change standards for radiation protection.

The debates shifted in the 1950s when the effects of nuclear fallout from weapons testing became public knowledge despite clumsy attempts by the military and political advisors in the Eisenhower administration to confuse the public with false stories. These included the claim that the Japanese fishermen on the *Lucky Dragon* in 1954 who were saturated with radioactive ash and rain and who showed symptoms of radiation sickness were actually victims of hepatitis. The AEC also claimed the fishing boat was actually on a spying mission to observe Project Bravo (the hydrogen bomb test in the Marshall Islands).[22] It included the claim that a little bit of radiation is actually good for humanity because heterozygous states tend to be more fit in an evolutionary setting than homozygous states (the sickle cell heterozygote resistance to severe illness from malarial infection was a major case used). There were also denials of stories in local papers that sheep and other livestock in the fields near Utah and Nevada, where atomic testing was frequently carried out, were dying and showed symptoms of radiation sickness or had given birth to malformed offspring. Nuclear fallout became a major contention in the Cold War debates as hydrogen bombs became larger and atmospheric testing in remote places became more common. At that time the U.S. had a double standard— using worldwide fallout detection to monitor and announce all Soviet H-bomb tests in Siberia and denial that worldwide fallout was of any harm. That assessment was a difficult one, because Barry Commoner and other scientists had initiated a "baby tooth" campaign, and showed that such teeth indeed contained radioactive elements from weapons testing and hence were a potential concern for bone cancers and blood cancers.[23]

At issue in the debates over low doses of radiation were two ways the same data could be used. Measured as individual risk, the low doses were of negligible concern. In Muller's estimates, only a minute elevation of risk (less than 1% over control rates) was present for radiation-induced leukemia for any child born to a parent in the U.S. But in Linus Pauling's assessment, that minute risk, multiplied by the world's population, resulted in several thousand unnecessary cases of leukemia (then almost all were fatal). Muller at first supported the government's unlimited nuclear testing, arguing that it was "better to be dead than red."[24] He disagreed with Edward Teller, who took the position that there was no harm at all from low doses of radiation, and if radiation at those levels had an effect, it was "to speed up evolution" and provide occasional beneficial mutations to improve humanity.[25] Muller shifted his views when the multimegaton hydrogen bombs on both sides of the Cold War began to produce significant amounts of worldwide fallout entering the food chain for humanity. He then called for and endorsed bilateral conferences to work out nuclear test ban treaties.

The National Academy of Sciences in 1955 set up a committee including Muller, George Beadle, Theodosius Dobzhansky, Milislav Demerec, Tracy Sonneborn, Alfred Sturtevant, and James Neel to prepare a report known as the BEAR Report (Biological Effects of Atomic Radiation). Most agreed that there was no safe dose for radiation, but they disagreed on when low doses of radiation became significant enough for government regulation.[26]

The success of the John F. Kennedy administration in eliminating atmospheric and underwater testing of nuclear weapons put an end to the intensity of the debate over fallout, but the issue of arms control remained active throughout the 1970s and 1980s and eased as the Cold War waned and both sides sought ways to reduce their stockpiles of weapons.

The medical community made progress in radiation protection. Practices such as using X rays to measure shoe size or to award students in contests for a perfect posture using X rays of spinal columns were prohibited. Physicians no longer use X rays to straighten out bowed legs in children. X rays are no longer used to shrink thymus glands (as was done in the 1950s) on the false premise that enlarged thymus glands lead to chronic upper respiratory infections. Physicians and dentists routinely use lead aprons on their patients when they receive diagnostic radiation. Physicians leave the room when they take X rays. Even more effective, however, were the advances in enhancing X-ray images and greatly reducing the doses of diagnostic radiation. Radiation safety is taught to health professionals, and strict standards are applied for use of radiation and for testing the machines that are used to administer X rays.

The nuclear industry at first had a lot in its favor. It promised cheap electricity with the prospect that the spent fuel could be recycled and even more fuel extracted than was initially used. The nuclear reactors did not emit carbon dioxide and other gases that contribute to global warming. They were cleaner and thus less likely to be a threat to health through lung cancers, emphysema, and other respiratory diseases. There were, unfortunately, some troubling aspects to this initial optimistic endorsement. Congress set limits on lawsuits in case of accidents. The nuclear industry was heavily subsidized for its research by federal funds (in contrast to alternative energy methods using sunlight, wind, tides, or geothermal sources). Most important, the agency in charge of regulating the industry was the same agency promoting its development and commercial success. This was as close to a blank check as a new industry could hope for. Its engineers produced a safety evaluation, the Rasmussen Report, which claimed a nuclear reactor had a chance of failing only once in 10,000 years.[27] It was not only circulated and trusted in the U.S., but it was used by Soviet engineers in their own promotion of the safety of their nuclear reactors.

The Three Mile Island disaster in Pennsylvania in 1979 and the Chernobyl disaster in the USSR in 1986 soured the public on the safety of nuclear reactors. The failure in both cases was both in engineering design and in human operation.[28] A report prepared by engineers is not as likely to take into account the potential for careless operation, bypassing of safety concerns, or the hubris that engineers are not so foolish that they could make such errors. Until those disasters, the nuclear industry bullied its critics as Luddites and ignoramuses. They were fully convinced that such accidents were virtually impossible because every engineering aspect had been carefully checked out.

From this brief overview of the Cold War issues and usage of radiation in society, we can identify several moral or ethical issues:

1. In times of war or the threat of war, it is difficult to have an objective debate because any criticism is filtered through a prism of politics.

2. The issue of one's position on low doses of radiation is not due to discordant data. It is the way that same data is interpreted from the perspective of the individual or the perspective of a global population.

3. National security is frequently invoked to justify deception, denial, or slander to protect a position favored by government advisors.

4. Scientists are vulnerable to overconfidence in their skills and play down their human tendency to make errors.

5. Some bad outcomes may arise through ignorance at the time that cannot be predicted.

6. Inadequate education can lead to abusive practices by those who only see the good of what they do.

These are not easy issues to address. The most important needs are openness to honest debate and the staffing of objective regulatory agencies independent of the people and institutions they regulate. Most difficult for scientists to regulate will be our own personalities. When Oppenheimer commented on his role after the bombings of Hiroshima and Nagasaki, he remarked, "Physicists have known sin."[29] Almost a generation later, in 1983, Teller denied Oppenheimer's self-incriminating statement. Instead, he claimed, "I would say that physicists have known power."[30] Is there a difference between sin and power? Oppenheimer was steeped in the values of his Ethical Culture upbringing as a youth and was troubled by his conscience. Teller did not wrestle with his values. Indeed, he pushed forward with the development of the more powerful H bomb. He knew what he was doing would not "save his soul," but he saw himself among those privileged to gain and use power. Teller's was a political universe ("real-politics" as the Germans designated it and as Secretary of State Henry Kissinger practiced it), and Oppenheimer's was still tinged with a moral dimension, inconsistent as it was.

Notes and References

1. J. Bergonie and L. Tribondeau, "Actions des rayons X sur le testicule du rat blanc." *Comptes Rendu Acad. Sci.* **143** (1907): 983; C.R. Bardeen, "Abnormal development of toad ova fertilized by spermatozoa exposed to Roentgen rays." *J. Exp. Zool.* **4** (1907): 1–44.
2. H.J. Muller, "The problem of genic modification. Proceedings of the Fifth International Congress of Genetics, Berlin, 1927." *Z. Induck. Abstammungs Vererbungsl., Supplement I* (1928): pp. 234–260. Lewis Stadler independently of Muller found X rays to induce mutations in barley and maize, his work appearing in 1928. L. Stadler, "Genetic effects of X-rays on maize." *Proc. Natl. Acad. Sci.* **14** (1928): 69–75.
3. G. Pontecorvo and H.J. Muller, "The lethality of dicentric chromosomes in *Drosophila*." *Genetics* **26** (1941): 165.
4. B. McClintock, "The fusion of broken ends of sister chromosomes following chromatid breakage at meiotic anaphase." *Research Bulletin of the Missouri Agricultural Research Station, Number 290* (1938).
5. H.J. Muller, "Analysis of the process of structural change in chromosomes of *Drosophila*." *J. Genet.* **40** (1940): 1–66.
6. S.P. Ray-Chaudhuri, "The validity of the Bunsen-Roscoe law in the production of mutations by radiation of extremely low intensity. Proceedings of the Seventh International Congress of Genetics." *J. Genet.* (suppl.) (1939): 246.
7. Many books have been written on the making of the atomic bomb. For the quotes in this chapter, I have used Peter Wyden, *Day One: Before Hiroshima and After* (Simon and Schuster, New York, 1984); Gregg Herken, *Brotherhood of the Bomb: The Tangled Lives and Loyalties of Robert Oppenheimer, Ernest Lawrence, and Edward Teller* (Henry Holt, New York, 2002). I also used John Beatty's appraisal (1991) "Genetics in the atomic age: The Atomic Bomb Casualty Commission, 1947–1956." In *The Expansion of American Biology* (ed. K.R. Benson et al.), (Rutgers University Press, New Brunswick, New Jersey, 1956), pp. 254–321.
8. Wyden, op. cit., p. 151.
9. Ibid., p.155.
10. Ibid., p. 161.
11. Herken, op.cit., p.132.

12. Wyden, op. cit., p. 159.
13. Ibid., p. 178.
14. Ibid., p. 213.
15. Herken, op. cit., p. 134.
16. Ibid., p. 134. Some critics of Oppenheimer have told me they felt his *Bhagavad Gita* quote was a sophisticated rationalization, and others told me it was a profound insight into the psyches of scientists applying their knowledge. I think both interpretations are valid. He was aware of the historical importance of the moment and he probably searched for an appropriate quote to make if the test was successful. Bainbridge's quote has spontaneity to it, and I don't think he prepared his remark in advance.
17. Ibid., p. 139.
18. Ibid., p. 142.
19. Ibid., p. 149.
20. See E.A. Carlson, *Genes, Radiation, and Society: The Life and Work of H. J. Muller* (Cornell University Press, Ithaca, 1984), pp. 352–367.
21. Willard F. Libby, interview with Elof Carlson, UCLA, Los Angeles, April 8, 1971 (Elof Carlson diaries, donated to Cold Spring Harbor Laboratory Library and Archives).
22. Beatty, op. cit., p. 289.
23. B. Commoner, "The fallout problem." *Science* **127** (1958): 1023–1026; also see Ernest Sternglass, *Secret Fallout: Low-level Radiation from Hiroshima to Three Mile Island* (McGraw Hill, New York, 1981).
24. Muller's views on radiation and weapons testing were complex, and he considered his interpretation superior to that of Teller (who denied harmful effects of low doses) and Pauling (who was very concerned with the global impact of fallout from weapons testing). Muller reasoned that there were two issues—one involving the risk of war and the other involving the risks of low doses of radiation. He felt the former were more serious, and thus, he opposed unilateral disarmament and favored weapons testing as a deterrent to the Soviet Union. He felt Pauling's concern stressed the harm of low doses from fallout and not the very real harm of actual nuclear warfare. He changed his mind in the early 1960s when hydrogen bombs in the 10–60 megaton range were being exploded; from that fallout he thought enough strontium-90 and other longer-lasting radioactive fallout would impair health.
25. Teller's views were expressed at a public lecture at UCLA that I attended about 1967.
26. For a discussion of the BEAR Report, see Paul Berg and Maxine Singer, *George Beadle: An Uncommon Farmer* (Cold Spring Harbor Laboratory Press, Cold Spring Harbor, New York, 2003). See Chapter 14 "Genetics and the nuclear age," pp. 223–239.
27. Norman Rasmussen, ed. 1975. *Policy on risk assessment: A policy statement of the American Nuclear Society. WASH-1400 of the Nuclear Regulatory Commission.* The report was issued in 11 volumes, of which volume 1 contains a summary of the Commission's report. Rasmussen did predict that accidents were possible through loss of coolant failure or through human error, but he was optimistic and estimated that risk as one failure in 10,000 years of operation.

28. In the Three Mile Island, Pennsylvania, accident, no one was injured or killed, but evacuation of the surrounding community was ordered by the governor of Pennsylvania. In Chernobyl in the Ukraine, there were many deaths (about 40) from radiation sickness of the operators and firemen. An ongoing concern is the contamination of soil with longer-lived isotopes as a plume of radioactive debris moved northward to Lapland and south to the Caucasus. Thyroid cancers, bone cancers, and leukemias are still being monitored in the areas of higher fallout dosage.
29. Herken, op. cit., p. 334.
30. Ibid., p. 334.

7

Herbicides in Peace and War

PLANT GROWTH IS REGULATED BY HORMONES. In the 1940s, Ezra Jacob Kraus (1885–1960), a botanist at the University of Chicago who was a pioneer in the early use of plant hormones, noted that some broad-leafed plants died when plant hormones caused their rapid growth.[1] The effect was dramatic, with leaves wilting and falling within 2 days after treatment. The compound he used was 2,4-dichlorophenoxyacetic acid (2,4-D). Kraus alerted the War Department about his findings, and the Army tested his compound but found no need to use 2,4-D in the last years of World War II. After the war, the railroads presented a commercial use for Kraus's herbicide. Spraying the weeds growing along the tracks with 2,4-D prevented fires and degradation of the tracks and made litter removal easier. The herbicide was also promoted by chemical companies for gardeners and ranchers who wanted to keep broad-leaf weeds from competing with grass.

Army chemists continued their experiments and found combinations of herbicides that were effective for clearing broad-leaf vegetation. Their most rapid and effective mixture was a combination of 2,4-D and 2,4,5-tricholorophenoxyacetic acid (2,4,5-T). At the time, the chemists were unaware that 2,4,5-T is contaminated with a product called dioxin, which arises at higher temperatures when 2,4,5-T is manufactured. A large number of dioxins form from the fusion of 2,4,5-T molecules, some of which are highly toxic. The most potent of these fusion products was 2,3,7,8-tetrachloro-dibenzo-P-dioxin, also known as TCDD. Dioxin readily binds to pores of the plasma membrane of cells and passes into cells of the skin.

In 1961, herbicides were sent to Vietnam. The military reasoning was that the Viet Cong who were fighting the war blended into the foliage and acted as snipers, shooting guns or launching mortars at U.S. and Allied boats in the rivers and in the mangrove forests. A second objective of the use of the defoliants was to starve the Viet Cong by spraying the farms that allegedly supplied their food. The herbicide was mixed as a thin paste and sprayed from planes in a program known as *Operation Ranch Hand.*[2] A number of mixtures of herbicides were used for different vegetation. They were identified on the drums that contained them by a colored stripe. Agent Orange was the most widely used; it was a mixture of 2,4-D and 2,4,5-T. Agent Blue was an arsenical herbicide. Other color-coded mixtures were chosen for specific crops and for other ecological sites.

Spraying was carried out by airplanes from 1961 to 1975 (with most of the spraying taking place in the period 1964–1971). About 72 million liters of herbicides were used on about 1.7 million hectares of South Vietnamese forests, amounting to about 10% of the landmass of South Vietnam. No spraying was done in North Vietnam. Agent Orange accounted for about 60% of herbicides used. Agent Blue was mainly used on cropland. Although most of the spraying hit the targeted areas, about 10% of the missions had to be aborted because of bad weather or enemy flak. In those cases, the planes would dump their contents, sometimes unintentionally drenching friendly villages or their own Allied (or U.S.) forces.

At issue in the debates about the use of herbicides in the Vietnam War are the following:

1. At the time the military approved use of herbicides on a large scale, the manufacturers knew there were health problems associated with the use of these products, but that information was kept secret from both the public and the government.[3]

2. Military observers were divided about the effectiveness of spraying. It did not diminish the infiltration of Viet Cong from the north to the south. It just changed their routes and shifted the locations where the fighting was carried out. Nor did it significantly cut off food supplies to the Viet Cong, because both the Allies and the Viet Cong were supplied by unsprayed farms and there was no way to identify which farms were supplying the Viet Cong. Despite those negative field assessments, the spraying program had a momentum of its own, and it is unclear what military objectives were satisfied.

3. Arsenicals do not degrade; thus, agricultural land sprayed with Agent Blue remained contaminated and some of the arsenicals ended up in the groundwater, drinking water, and food chain after the war.

4. 2,4-D and 2,4,5-T rapidly degrade on exposure to the sun, and most of these herbicides were gone within 2 years after the war ended. But

dioxin has a half-life of 10 years in soil, and thus, significant amounts were present for at least two generations after the war ended.

5. A significant number of American veterans were heavily exposed. Those in the mangrove forest areas had soaked up to their waists in waters heavily laced with herbicides. The spraying often created mists that drifted many miles from the target area and descended on U.S. troops. Despite this exposure, the government (in both Republican and Democratic administrations) refused to acknowledge that a health hazard existed.

6. By far, the most exposed people were the Vietnamese who had to farm (during and after the war) in the sprayed areas or on contaminated soil and who had to live out their lives in ecologically damaged parts of the country. There were no ways to decontaminate the sprayed areas. It was assumed that natural degradation of the chemicals would take place. It was also assumed that herbicides (other than the arsenicals) are safe and posed no health hazards.

7. No serious thought was given to the ecological effects on the flora and fauna of the mangrove forests or the sprayed nonfarming areas. Forests are diverse and complex in their interaction. Removal of canopies from forests and deforestation of habitats have consequences on the ranges for different plant and animal species. The assumption most favored was that nature would repair itself when the war ended.

Biological Effects of Herbicides

My own interest in these herbicides arose when I was contacted by a genetic counselor at my university. She had been asked by a congressional representative in Nassau County, New York, to supply him with information he could use for veterans who inquired about health hazards of Agent Orange. I told her I would read the literature and let her know what I found out. I had earlier had an interest in environmental mutagens because I had tested lysergic acid diethylamide-25 (LSD) when I was at UCLA and found that it did not produce gene mutations, chromosomal rearrangements, or increases in nondisjunction when tested on fruit flies (by injection of a dose estimated at 750 times a human trip dose). At that time, there was a conflicting literature on LSD as a mutagen, based mostly on the new field of tissue culture cytogenetics. I found the same to be true for the literature on Agent Orange. Since I knew how to test for environmental mutagens in fruit flies, I did a series of experiments with my students on 2,4-D and 2,4,5-T, as well as an Agent Orange mixture. I used a dirty sample of 2,4,5-T that a colleague had from Dow Chemical that should have been high in dioxin content (it was in the batches made before the dioxin concern led to reduced temperatures during manufacture). We did not find any genetic effects but we did find the following[4]:

1. Both 2,4,D and 2,4,5-T caused females to lay fewer eggs on the petri dishes containing food.

2. There was a developmental delay for both sexes. Normally, males and females emerge from their pupa cases about 10 days after eggs are laid. The herbicides delayed this eclosion process, with females seriously delayed (also all the first emergent flies were males).

3. There was some teratogenic effect on the development of the flies because many of the emerging adults had everted genitalia (possibly caused by protrusion of the gut).

4. At high doses, the eggs laid failed to develop beyond the first or second larval stages.

5. A common cause of death among the larvae involved premature migration of second-instar larvae to dry places on the sides of the bottle, where they would dry out and die.

One reason for this sex difference in response to herbicides is that both 2,4-D and 2,4,5-T are readily soluble in lipids. Because female fruit fly adults manufacture large numbers of eggs amounting to several times their body mass, and because these eggs are lipid-rich, there is more herbicide stored in those tissues than in the males.

One cannot extrapolate with confidence from animal experiments to human consequences, but it is important here (as in the animal studies required by law for FDA-approved prescription drugs) to note that there are significant biological effects of these herbicides on fruit flies, and a prudent manufacturer would order more tests on mammals. It would also be reasonable to look at lipid-rich tissue as the site for storage of dioxins and herbicides after exposure. This would include the skin and the nervous system as well as the plasma membranes of cells. The effects of these herbicides are not limited to plant life alone.

Health Effects of Agent Orange

The federal government for a long time refused to acknowledge that there were health effects of herbicides on U.S. veterans. The only effect they acknowledged was an acne-like rash called chloracne that appeared on the skin of exposed soldiers. Other than the esthetic unpleasantness of chloracne, the government did not consider the herbicides to be harmful. This policy is consistent with governments who do not wish to pay large sums to care for the chronic health problems or delayed health problems of veterans. It was the same reluctance used for veterans exposed to radioactive fallout during the early years of the Cold War when atomic tests were carried out in Nevada. To be sure, old people, exposed or not, will have their share of cancers, psychological depression, impotence, and chronic pain.

The problem for federal agencies responsible for veterans is that the government has always been wary of cheaters who claim to have been exposed so they could get medical benefits for their ailments. Chloracne is different. You can see it, it is not common in unexposed adults, and eventually it goes away or becomes less troublesome. If some 20% of the population will eventually experience a cancer, and if the Agent Orange-exposed have a slightly increased frequency of cancers, the government will surely have all veterans who develop cancer later in life believing that their cancers were caused by Agent Orange.

I attended the *International Symposium on Herbicides Used in the Vietnam War* which was held in Ho Chi Minh City, January 13–20, 1983.[5] The meeting was organized unofficially by Cyrus Vance through the United Nations (at the time the U.S. still did not recognize the Vietnamese government). The Vietnamese presented their evidence for ecological and health damage from the use of herbicides during the war. For the health damage they offered the following:

1. An increase in birth defects among the populations in the sprayed areas of the south and among the children of returning Viet Cong who raised their new families in the unsprayed north.

2. An increase in liver cancers among the persons sprayed.

3. An increase in chromosome breaks among the cells of those exposed to herbicides during the war.

4. An increase in molar pregnancies (more than 100-fold) (see below for

more information) among the women living in the sprayed areas but not among the pregnancies of returning veterans to the north.

During the sessions discussing these presentations, the American and European delegates raised a number of objections to the findings or sought clarification. These included the following:

1. The ascertainment of birth defects was unclear. Most of the birth defects were self-reported by parents when visited by health-care workers (equivalent to the barefoot doctors program in rural China). Diagnosis was difficult because of the poverty of the country and the lack of access to urban clinics. When I pointed out that their unexposed control population had a birth defect incidence far lower than that found in the U.S. or Europe, they replied that they were an agrarian and not a manufacturing country, so they had fewer pollutants. That is a dubious claim, and one could just as readily claim that parents with a child with a birth defect were more likely to claim or guess that they were in an Agent Orange-exposed area than to claim they had crummy genes. That would also result in an unnaturally low spontaneous birth defect frequency.

2. The use of chromosome breakage, sister chromatid exchanges (similar to somatic crossing-over), and deletions was done with standard acetocarmine dyes rather than Giemsa or quinacrine staining. That is, they did not use chromosome banding to locate breaks or regions of exchange or loss. This was the same problem with the early LSD studies in the U.S. Many of the alleged breakage events may have been the artifacts of procedures for squashing, dyeing, and handling the cells in tissue culture. They did not have the microscopes, the funding, or the trained personnel to do cytogenetic studies at a level of quality of a genetic services laboratory in Europe or the U.S.

3. The liver cancers may have been Agent Orange-associated as U.S. veterans also claimed. But in Vietnam there was a higher incidence of hepatitis C, which is associated with a higher incidence of liver cancer, and there were no effective controls to rule this out in the Vietnamese population.

There was only one piece of evidence that I felt was suggestive of a health problem, and I had no explanation until a few years later when I read an article about the origin of molar pregnancies. These are found rarely in the U.S. (about 1 in 1000 pregnancies) and come in two types.

One produces extraembryonic membranes, but there is no embryo inside to develop. It is otherwise benign. The second type also lacks an embryo but is associated with a cancer in the chorionic tissue that can be lethal to the pregnant woman. The article noted that a molar pregnancy (in mice) arises when an egg expels its own nucleus and only has a sperm nucleus for its nuclear chromosomes. This androgenetic zygote usually doubles its haploid male content. If the resulting zygote is XX, it forms the benign molar pregnancy; and if a YY zygote forms, it results in the cancer-prone molar pregnancy.[6] The Vietnamese reported rates of molar pregnancy of about 5–10%, which would be far in excess of an ascertainment bias. One possible reason for this excess is that the eggs of women in the sprayed areas take in more dioxin and this might cause the nucleus to be extruded. When the egg is fertilized, it results in the androgenetic zygote and a molar pregnancy. As far as I know, no one has explored this possibility in Vietnam. It would not be significant for U.S. veterans, because virtually all of them were male during the Vietnam War.

Because of Cold War politics and the failure to recognize the Vietnamese government, both our own veterans and the Vietnamese people lost immense opportunities to study the health effects of herbicide spraying. One could have plotted the birth defects and molar pregnancies by year from the end of the war to look for a drop in incidence paralleling the decay of the herbicides and dioxins in the soil. Similar prospective studies could have been done for soft-tissue cancers and for neurological ailments.

Ecological Effects of Herbicides

While I was at the symposium in Ho Chi Minh City, we were taken on excursions to some of the areas that were heavily sprayed. This included the Ma Da forest area, located between two rivers; the spraying resulted in a denuding of the area that cut the habitat in half. The dead trees were replaced by Imperata grass (*Imperata cylindrica*), which dries out in the summer and readily catches fire. Some farmers used a slash-and-burn strategy and thought they could clear the ground of stumps and use the soil for commercial crops. Unfortunately, jungle soil is shallow, often clay-like, and does not support most crop growth. The Vietnamese abandoned their pineapple, fruit orchard, bamboo, and vegetable garden projects. The natural fires (mostly from lightning) burned the dried grass that favored the soil and actually enlarged the divide, because fires would burn some of the boundary jungle next to the Agent Orange-framed divide.

Vietnamese biologists told us that many species were endangered and many had disappeared from these sprayed areas. The jungle canopy had not returned (this was 15 years after the spraying had ended), and it had provided a habitat for birds that are no longer present. Most hard-hit were the mangrove forests (which another group toured for the U.S. National Academy of Sciences), where both the vegetation was destroyed and the fisheries seriously damaged. At the time of the tour, the mangrove forests had not returned, and some of the ecologists estimated it might take a century for its return to prewar conditions.

Although this is a somewhat bleak ecological portrait, I do want to put it in perspective. Most of Vietnam was not sprayed, and on the occasion I later visited it (in 2001) while teaching on board the *SS Universe-Explorer* for *Semester at Sea*, I took a trip with some of the students along the Mekong Delta, and it was a luxuriant jungle growth showing no evident damage from the wars some 25 years earlier. I was also impressed by one Vietnamese resident in Ho Chi Minh City who listened to some of the students discussing Cold War politics and reminded them of his perspective, "Vietnam is a country, not a war."

Some Ethical Reflections on the Herbicide Controversy

There are a number of concerns about the civilian and military use of herbicides 2,4-D and 2,4,5-T.

1. Dow Chemical knew of health effects of its herbicides since February 22, 1965, when several of its scientists met to discuss the hazards of dioxin in 2,4,5-T production. They widened their concern when they met a month later with Monsanto and Hooker Chemical companies to pool their knowledge about TCDD (dioxin). In June, 1965, they described (for internal use) TCDD as an agent that produced chloracne and "systemic injury." Dow had conducted a study in rabbits and showed that dioxin produced severe liver damage. They were concerned that the government might learn of their findings and the chemical industry might suffer financial losses as a result. Despite that concern, both the chemical manufacturers and the military insisted that the defoliation was a military necessity and that herbicides presented no serious health problems to the army or to the civilian population sprayed.[7]

2. The military knew as early as 1967 that herbicides had health consequences on those exposed to it. The author of the history of Operation
Ranch Hand for the army, James Clary, an Air Force scientist, informed
Congress in 1988 that "We were aware of the potential for damage due
to dioxin contamination in the herbicide. We were even aware that the
'military' formulation had a higher dioxin concentration than the
'civilian' version, due to the lower cost and speed of manufacture.
However, because the material was to be used on the 'enemy,' none of
us were overly concerned."[8]

3. The military used a deceptive message to reassure the press about concerns over Agent Orange use: "The purpose of this exercise would be
to meet criticisms of excessive use of defoliants by clarifying that they
will no longer be used in large areas, while in reality not restricting our
use of defoliants (since they are not now normally used in this area
anyway). In addition, there would be an escape clause ... which would
permit the use of defoliants even in the prohibited area provided that
a strong case could be made..."[9]

4. In the early 1960s, about 50 factory workers who manufactured Agent
Orange for Diamond Alkali Company in New Jersey were treated for
ailments, especially "painful and disfiguring" skin diseases (probably
chloracne). Although the company alerted the state health officials,
nothing ensued because the war production of herbicides took prece-
dence over the employees' complaints.

5. Congressional hearings that led to a veteran's benefit act in 1991 were
chaired by Congressman Ted Weiss (Democrat, New York). He had an
unusual ally in Admiral Elmo Zumwalt, Jr., who had supervised much of
the spraying program. Zumwalt's son was exposed to Agent Orange and
died young of a cancer that Admiral Zumwalt felt was induced by herbicides during the war. Among Weiss's committee findings, he reported
"the only thorough study of workers exposed to dioxin, released last year
by the National Institute of Occupational Health and Safety found that
the workers receiving high exposure had a 46 percent excess of cancers.
This is hardly modest."[10] Weiss accused the federal government of both
a cover-up policy and deliberate deception to shield the agencies and the
chemical industries from expensive litigation.

Had there been no Operation Ranch Hand, there would have been a
more concerted effort to hold chemical companies liable for health prob-

lems arising from herbicide exposure. As was true for the development of the atomic bomb, in times of war or national emergency, there is a tendency for values and justice to take a back seat to patriotism. What is troubling about the logic of Operation Ranch Hand is the military assessment that designed it and maintained it against the logic (that turned out to be correct) that an irregular army of Viet Cong, some in civilian garb, would fight a guerilla warfare using flexibility as their strategy. If they were thwarted by defoliation in one region, they would go through jungles unseen in an adjacent unsprayed area. It was impossible to defoliate all of south Vietnam without a calamity befalling the Vietnamese who supported the U.S. presence and the American soldiers who depended on food and supplies raised in South Vietnam.

Note that throughout this analysis of the historical circumstances of herbicide use, utilitarian ethics are used to justify military programs for defoliation. The decisions to withhold information on health damage are based on principles of loyalty (to one's peers or stockholders). The decisions to hold off veteran's benefits for Agent Orange disorders (real or alleged) are based on a utilitarian principle that the costs outweigh the benefits and a hidden assumption that veterans complaining about damage are really alcoholics, cranks, and cheats who are using Agent Orange as a pretext to make the government pay for their personal failures.[11] In a democracy, compromise is the usual outcome of controversy, and the veterans got some belated help for their illnesses. To my knowledge, the millions of sprayed Vietnamese (a considerable portion of them our allies during the Vietnam War) got nothing. One can call that bad outcome a post-casualty of war.

Notes and References

1. E.J. Kraus and J.W. Mitchell, "Growth-regulating substances as herbicides." *Bot. Gaz.* **108** (1947): 301–350. Kraus was one of four groups that independently used or synthesized 2,4-D and 2,4,5-T in the early 1940s but did not get to publish their results until after the war because of secrecy policy governing applied science during the war years in Britain and the U.S. See James R. Troyer, "In the beginning: The multiple discovery of the first hormone herbicides." *Weed Sci.* **49** (2001): 290–297.
2. William A. Buckingham, Jr., *Operation Ranch Hand: The Air Force and Herbicides in Southeast Asia, 1961-1971* (U.S. Government Printing Office, Washington DC, 1982).
3. Anonymous. *U.S. Veterans Dispatch Report, November 1990,* 24 pp. See page 2. This report is also available on the Web [www.usvetdsp.com].
4. E.A. Carlson, D.A. Ciccarone, G.D. Jay, L.J. Moss, R.S. Levy, and T.K. Myers. "Effets

biologiques des phenoxyherbicides sur le *Drosophila melanogaster*." In *Symposium International sur les Herbicides et Defoliants Employés dans la Guerre: Les Effets a Long Terme sur l'Homme et la Nature*, Volume III, pp. 190–199. Comité national d'investigation des conséquences de la guerre chemique EUA au Vietnam, Hanoi. The work on LSD is in Dale Grace, Elof Axel Carlson, and Philip Goodman, "*Drosophila melanogaster* treated with LSD: Absence of mutations and chromosome breakage." *Science* **161** (1966): 694–696.

5. The symposium was published in about 15 volumes, all the articles translated into French (see footnote 4).

6. In humans there are no YY androgenetic molar pregnancies. More than 90% are 46, XX, and the sperm chromosomes double after entry into the enucleated egg. The other 10% are XY with both sets of chromosomes derived from sperm (an entry of two sperm into the enucleated egg). Partial molar pregnancies also occur less frequently and they are often triploid, involving two or three sperm entering the enucleated egg. In a partial molar pregnancy, there is disorganized fetal tissue. For the human status of the androgenetic origin of molar pregnancies see N.G. Wolf and J.M. Lage, "Genetic analysis of gestational trophoblastic disease: A review." *Seminars in Oncology* **22** (1995): 113–120.

7. Anonymous op. cit., pp. 3–4.

8. Ibid., p. 4.

9. Ibid., p. 5.

10. Ted Weiss (1927–1993) represented the West Side of Manhattan and chaired committees on veterans' health care. He was an army veteran who served in 1946 in Hiroshima and took an interest in environmental issues. He received a Legislator of the Year Award in 1991 from the Vietnam Veterans of America for getting through Congress an Agent Orange Bill, HR 556 (January 29, 1991). It extended disability and death benefits to those exposed veterans with non-Hodgkin's lymphoma and soft-tissue sarcoma. I knew Ted Weiss because he was married to my wife's sister, but we had agreed early after his election to Congress that we would respect each other's commitments to our careers and not impose on each other.

11. I used to talk about Agent Orange to veterans groups, and one Veterans Administration representative told me privately that I should not be duped into believing most of the veterans who spoke out about their disabilities because they are more likely to be alcoholics and psychotics. I did not believe, and had no way of knowing whether, those who spoke to me were psychotic or alcoholic. They were certainly articulate in describing their health problems (inability to focus; tiredness; loss of interest in their relations to others or jobs). Whether dioxins can persist in neurons or alter their function more or less permanently would be difficult to demonstrate.

Part 4

THE REGULATION OF SCIENCE TO PROTECT INDIVIDUAL AND PUBLIC HEALTH

O NE OF MY COLLEAGUES IN MEDICAL SCHOOL once complained to me that regulations on drug testing in his field (psychiatry) took so long to get approved that his field was severely impaired. He felt that the thalidomide disaster in the 1960s was a bizarre oddity and not at all possible today even without government regulation. I disagreed, because without regulation the tendency to cut costs, compete for an early introduction of a new product that is immensely profitable, and accept shoddy testing is too great a risk. Thalidomide abuse led to 8000 known deformed babies and probably an even greater number of aborted embryos that were dismissed as spontaneous miscarriages. I chose the thalidomide case because it was not, as I had thought originally before I read in detail about its history, an "act of God," but a cautionary tale about self-deception, carelessness, and dishonesty. It is also the story of courage, by Frances Kelsey, in her efforts to resist both political pressure and the persistent efforts of a drug company that was impatient with the requests made for more testing.

The diethylstilbestrol (DES) story is different. No deception was involved, but the issues raised are not totally resolved, even today. The quality of research in medicine is not as carefully done as in those life sciences where human applications are remote. There are many reasons for this. The costs are high; the sample size of tested individuals is small; and the training of physician scientists may not be as intense and demanding as it is for those who receive Ph.D.s. The discredited or unproved idea of

95

"hormonal insufficiency" as the cause of most spontaneous miscarriages has once again emerged, with clinicians urging the use of hormones to prevent them. One would hope that solid experimental evidence and not anecdotal cases would be used to justify such a revival.

I did not choose current controversies over Vioxx (allegedly a cause of heart disease for arthritics using this product as a painkiller) or Viagra (allegedly a cause of blindness for some using this product as a treatment for male impotence) because the stories are still unfolding on these, and it is difficult to assess charges and rebuttals when the number of reported bad side effects is very low (compared to the high incidence of bad outcomes in the thalidomide and DES cases).

Striking a balance between the patient's need for more products, and that same patient's need for products that won't do more harm than good to an unlucky few, is difficult. Although drug companies and their lobbyists claim the high cost of drugs is caused by regulations imposed by the Food and Drug Administration in the U.S., many economic analyses of the price of drugs show that marketing costs are the major factors in these high costs. Very clearly, as anyone watching the evening news will experience, prescription drugs are aggressively marketed at a very high cost to influence physicians to prescribe them and patients to demand them.

8

Thalidomide—Corporate Misconduct Masquerading as an Act of God

THE "ACT OF GOD" MYTH IS A COMFORTING STORY. In this mythic form, thalidomide was a prescription drug tested as well as it could be by the standards of the 1950s in Germany. It was allegedly designed and used to treat morning sickness. When malformed babies became an epidemic, Chemie Grünenthal, so the myth goes, immediately pulled thalidomide off the market. Both its manufacturer and the governments of the places where these babies were born allegedly contributed funds to their care, although the cause of the disaster was called by them an act of God. Virtually every statement in this myth is false.[1]

Thalidomide was indeed synthesized in 1954 in Germany by Wilhelm Kunz while he was studying ways to produce peptides. However, its manufacturer, referred to in the shortened form as Grünenthal, was looking for a sedative to compete with Miltown and other successful tranquilizing drugs in the U.S. Kunz's colleague, Herbert Keller, thought it looked like a barbiturate and thus took over its pharmacological analysis. It was not initially considered for the treatment of morning sickness. Grünenthal used standard tests on rats and mice but couldn't demonstrate that their drug had a sedative effect. They had to invent a new test to claim thalidomide was effective. They tested it for addictiveness and found it wasn't addictive, so they asked for and got permission to market thalidomide, not by prescription, but as an over-the-counter drug. At first, it was used as originally intended for adults who needed a sedative. Later, it was used for women with morning sickness. No tests were done on embryos to study its effects on pregnant women using it for morning sickness. Side effects for its use as a sedative were reported and ignored or dismissed. When thalidomide was suspected of causing birth defects, Grünenthal tried to intimidate physicians by threatening to sue them for slander. When thalidomide was finally proven to be the cause of those birth defects, Grünenthal reluctantly withdrew the drug. Grünenthal used a fleet of lawyers to block lawsuits, apply gag rules, and delay (up to 15 years) settlements. Grünenthal used the "act of God" myth to lobby government legislatures for funds for thalidomide children. Someone at Grünenthal may have shredded its records (or they mysteriously disappeared) before it was sued. No criminal cases arose from the manufacturer's hardly acceptable practices.

The legacy of the thalidomide case is ongoing. Most countries have strengthened or enacted strong testing procedures for new drugs to be marketed. The cost of drugs is much higher because those testing costs are absorbed in the price of the medication. It takes longer to get a drug onto the market, and some have argued that there are fewer new drugs on the market because of the inhibitory costs, risk liabilities, and regulations that stemmed from this failure. By playing it safer, so these critics claim, we are deprived of more effective drugs.

What Is Thalidomide?

Thalidomide is an organic compound consisting of three asymmetrical rings and the intimidating biochemical name of 2-(2,6-dioxo-3-piperidinyl)-1*H*-isoindole-1,3(2*H*)-dione; it resembles known barbiturates in structure. When it is manufactured, it comes in two mirror-image isomers or forms (an S and an R form). The R form is damaging to embryonic development. The S form is not. But in the human body, pH changes will convert the S form into a mixture of S and R molecules, and thus, there is no safe (nondamaging to embryos) form that can be used. This is important today because thalidomide has been reintroduced as apparently effective for a variety of disorders (especially leprosy, certain cancers, and immune diseases).[2] It was first introduced on the market in Germany on October 1, 1957. Eventually it was sold in 46 countries under a variety of trade names. In Germany it was called Contergan and in Great Britain it was called Distaval. It was not approved by the U.S. Food and Drug Administration because the official in charge, Frances Kelsey, did not think it had been proven to act as a sedative and she was concerned by reports of side effects. Its mode of action is still unclear. Among the probable functions assigned to it is that it acts as a suppressor of tumor necrosis factor-α, and this might explain some of its effects on the immune system and on certain cancers (multiple myeloma). It also inhibits vascular growth, and this might be another reason for its effectiveness on tumors. One of its effects on embryos seems to be associated with a failure of neuronal development in dorsal root ganglia.[3]

What Are Teratogens and How Do They Work?

A teratogen is a substance that causes damage to a developing embryo. Some teratogens do so by inducing mutations or breaking chromosomes. X rays and atomic particles fall into that category. Thalidomide is a teratogen that probably works in a different way. It has not been studied intensively for gene mutations or chromosome breaks, but the damage it does to embryos differs profoundly from damage to embryos exposed to radiation. In humans, fertilization marks day 1 of pregnancy. The embryonic mass implants about day 7, and it forms organs for the next 7 weeks. Hence, the first 55 days of pregnancy are described as organogenesis. At the end of this time, the human embryo becomes a fetus. No new organs are formed, but the organs formed during organogenesis enlarge and become more complex in structure. The first 3 months of human pregnancy are called the first trimester. Fetal development is rapid in the second trimester. By the third trimester, the fetus can survive early birth, especially with medical help. Second-trimester fetuses (4th and 5th month) rarely survive after premature birth, although a few 5-month-old fetuses manage to survive (often with medical problems) when given heroic medical attention. Agents that are harmful to embryonic development do their major damage in the first trimester (during organogenesis). Thalidomide has little known effect on fetal development. Because morning sickness occurs during the first trimester (usually 30–60 days or so after fertilization), the major effects of thalidomide were at the most vulnerable time for organ formation, when limb buds, facial features, and genitalia are being put together. The heart and central nervous system are formed earlier in organogenesis. Teratogens may act by interfering with cell division, by blocking embryonic chemical signals, or by cutting off the blood supply to new tissue.

Thalidomide was so potent a teratogen that women who took a single pill during organogenesis had children with birth defects. The defects caused by thalidomide include malformations of the limbs, ears, eyes, and face, and a variety of internal malformations involving the intestines, kidneys, ureters, and bladder. Additional effects on sexual development affected the penis, scrotum, uterus, oviducts, and vagina. Many women aborted (miscarried) or had stillborn infants, but records of those are more difficult to obtain. Miscarriages in the first month would not have been recognized as associated with thalidomide (they would have been interpreted as a skipped period). There should have been increases in 2nd- and 3rd-month miscarriages, but these are fairly common anyway (about one in five women will experience a miscarriage of a known pregnancy in her reproductive lifetime), and they would not have resulted in lawsuits or complaints by physicians passed on to Grünenthal because the women or their physicians would have assumed they experienced a spontaneous abortion.

Limb malformations were the major novelty of thalidomide use. The total absence of limbs is called amelia. The reduction of limbs to a flipper-like appearance is called phocomelia. Sometimes the ulna or radius of the arms would be distorted, instead of simply failing to grow. In general, the time when thalidomide was first taken would determine when in organogenesis the damage began. Similarly, discontinuing thalidomide in the first trimester would determine when the damage would stop. For these reasons, some babies showed multiple defects of limb, face, and internal organs and others were limited to defects of either the arms or legs. Because morning sickness does not occur after the first trimester, little is known about the use of thalidomide on fetal development in humans.

What Were the Motivations of Those Who Developed Thalidomide?

At Grünenthal, a major motivation was to find a sedative that could successfully compete with the very popular and profitable sedatives developed in the U.S. Exactly how effective thalidomide was as a sedative is not known because the standard tests based on rats in exercise wheels or "righting reflexes" (how soon rats get up after they are tipped over) were unsuccessful when thalidomide was tested. Instead, Grünenthal scientists developed a "jiggle box" in which the vertices of a platform were attached to electrodes, and as the rats moved about, any electrode in contact with a salt solution would be registered.[4] This way, the mean activity of a given number of rats could be measured for control and for thalidomide-fed rats. On this basis, Grünenthal claimed thalidomide was a sedative. Some safety testing procedures were of concern. When thalidomide was used as syrup on rats and dogs, some died.[5] They did not die if the thalidomide was supplied in a water solution. However, one feature was particularly useful for marketing thalidomide; it had no addictive effects, unlike typical barbiturates.

The pharmaceutical industry is competitive, and companies do have to worry about bad publicity over their products. When thalidomide was first used as an over-the-counter sedative, most of its purchasers were adults who were stressed out. Some complained of prolonged constipation after using thalidomide. Others complained of a condition called peripheral neuritis, in which the hands and feet tingle or become numb and lose the sensation of touch. Grünenthal's staff issued a form letter to physicians who wrote in or to persons who complained directly to the company.[6] The letter claimed that the product had been sold to millions of satisfied users, and theirs was one of very few complaints they had received.

Grünenthal was convinced it had tested the product to the satisfaction of regulations and law then in place in Germany. It was satisfied it had a product that worked, that was harmless, that was not addictive, and that would take over a substantial share of the world's market. They leased their production rights to Distillers Limited in Great Britain for distribution in the British Commonwealth, and they tried marketing it in many other countries. In most of their efforts they were successful. But they failed in the U.S.

The American Refusal to Grant a License for Thalidomide Use

Several American companies were contacted by Grünenthal regarding thalidomide. Two companies rejected the drug because they did not like the jiggle box evidence that the product was a sedative.[7] The U.S. Food and Drug Administration (FDA) requires that a product does what it claims to do and that it is safe for those who use it. One company, Richardson Merrill in Cincinnati, was interested in obtaining sales rights from Grünenthal, and they worked out an arrangement for some local physicians to give the product to their patients. At this time, thalidomide was being marketed in Europe for treatment of morning sickness. The data obtained from the physicians were analyzed by the staff at Merrill, and they assigned staff members to write up the articles and place the physicians as authors (the ghostwriters' names did not appear on the articles).[8] They sent these articles and their request for a license to the FDA for approval. Frances Oldham Kelsey was assigned this case. She had a cautious attitude because she was aware of a disastrous incident in 1937 when 105 children and adults died in the U.S. from use of sulfanilamide turned into a tasty elixir with antifreeze (diethylene glycol) as a solvent. It is astonishing to know that until 1938 there was no regulation that required drug manufacturers to test their products for toxicity! The FDA was established that year as a congressional response to public outrage over the deaths of these people, who believed they were being treated for an infection and instead had an "elixir of death" destroy their kidneys.[9] Dr. Kelsey read reports of the claims of peripheral neuritis and reviewed the evidence that thalidomide was a sedative. She believed Grünenthal had not demonstrated the drug's efficacy and she wanted more tests to study the claims of peripheral neuritis. She rejected the application.

The Richardson Merrill staff was furious, and they accused Kelsey of being "an officious bureaucrat." Over the course of a year, they called her office or sent letters amounting to 50 different requests to expedite the licensing of the drug. They accused her of being libelous in her responses to their inquiries. They wanted her fired or their application reviewed by someone less biased. Kelsey remained firm in her refusal and had the confidence of her immediate supervisors.[10]

The Widespread Epidemic of Phocomelic Infants

Grünenthal's investigators noted that nursing mothers who were given thalidomide as a sedative did not have sleepier babies.[11] They assumed that if thalidomide did not pass from breast milk to infant, then thalidomide would not be transmitted from the mother to the developing embryo or fetus. On this indirect evidence, they recommended that thalidomide also be used as an anti-nausea medication for women complaining of morning sickness. It is not unusual for drug companies to extend the uses of a product, but it is unusual to make such claims without lengthier trials to see that it is both effective and safe. One reason Grünenthal's staff may have had an illusion of safety was the absence of any addictive effects and the relative absence of toxicity of thalidomide in animal tests.

The first cases of phocomelia that began to appear in Germany were surprising because phocomelia was so rare that most hospitals had not encountered prior cases. The presence of a cluster of such cases caused concern among obstetricians. Similar reports began to appear in other countries. In Australia, it was first thought that radiation might be involved, because several of the women giving birth to such children lived near a nuclear reactor.[12] But a careful study of the women giving birth to these children in Australia found one common factor. They had all received thalidomide for control of morning sickness, according to Dr. W.G. MacBride, who made the analysis. A similar observation was made by Dr. Widukind Lenz in Germany. He urged Grünenthal to recall thalidomide. Grünenthal's lawyers threatened to sue him for libel, and they made sure the press knew that Lenz was the son of one of the architects of Nazi race hygiene during the war years.[13] Unfortunately for Grünenthal, the cases began to appear in large numbers, hundreds of babies with deformed limbs appearing as an epidemic in Europe and other countries where thalidomide was marketed. It was absent in countries that had not sold it. About 8000 babies with these birth defects were born. Many thousands more probably aborted early or were registered as miscarriages.

The Failed Attempts for Patient Compensation

In Great Britain, details of a case cannot be discussed in public before a verdict is rendered. This was fortunate for Distillers because their many lawyers used every available means to delay the cases. It took 15 years to work out a settlement, and many of the parents were forced to sign away their rights for a much smaller sum, which they desperately needed for the problems their children faced. Class-action suits are rare in most countries, and thus, most litigation involved one lawyer for the parents versus dozens or hundreds of lawyers for the defendant company. Other countries had statutes of limitations that favored the supplying companies, and the cases could not be filed or were dismissed. Grünenthal used its lobbying efforts to convince the German government that this was an act of God and not deliberate negligence on their part so that the taxpayers rather than the company should pay compensation to the families. Not a single case went to a jury for a verdict in any country.[14] Despite some investigative reporting that suggested corporate misconduct, no one was held responsible for the tragedy, and no one could explain how the pertinent documents managed to disappear.[15] The one country where a settlement was worked out between the families and the supplier was Great Britain. A compromise was agreed on some 15 years after the children were born. As is true of most commercial lawsuits involving a settlement, the two parties are forbidden to discuss how much was given to each family.

The Medical Treatment and Care of Thalidomide Children

The birth defects of thalidomide children vary in intensity and kind. In many cases, tags of flesh that replaced ears, and small boneless stumps without digits for arms or legs, were removed for aesthetic reasons. This was sometimes a mistake because it made prostheses more difficult to attach. The assumption used for plastic surgery was that the child would do better psychologically with some sort of symmetry that was less frightening to others who saw the child, especially other children. The children were remarkably adaptable and learned to use those stumps, flippers, and deformed arms and legs. They required a lot of training to get dressed and to care for their personal hygiene.

Many of these children, when they became adults, formed a thalidomide awareness organization, and they have fought hard to prevent the reuse of thalidomide as a prescription drug.[16] They believe that errors will happen, adults will be careless, and some pregnant women will eventually be exposed to its teratogenic effects. Despite sympathy for their concern, there are strong advocates for the carefully regulated prescription of thalidomide for a variety of disorders for which other treatments are ineffective or not available. This includes special instruction about the use and safe storage and disposal of thalidomide.[17] The FDA, for example, approved the use of thalidomide for treatment of the lesions of leprosy (Hansen's disease) in 1998. They also instituted an oversight program to ensure that no pregnant women would use the drug.

Lessons Learned from the Thalidomide Tragedy

The most important consequence of the thalidomide tragedy was a reform of drug-testing laws in most countries. It takes longer to get drugs through the research phases for efficacy and safety. In most cases, 10 years is required to go through the various animal and human trials before a license is granted. This makes the costs of drugs higher (although the highest costs for new drugs involve their marketing), and it frustrates both physicians and patients who have to wait for new products to enter the market. They would prefer faster trials and less elaborate procedures to speed-up the availability of promising drugs for depression and other psychoses, as well as life-saving drugs for cancers and chronic organ diseases.[18]

In 2004 in the U.S., the FDA was once again in the news. Several drugs with harmful side effects were pulled from the market, despite initial efforts by the drug companies to override the views of the FDA staff that consider these products worrisome. These include the COX-2-selective nonsteroidal anti-inflammatory drugs Vioxx (rofecoxib) and Bextra (valdecoxib). Another similar drug, Celebrex (celecoxib) is under FDA investigation. Labeling changes regarding drugs of this type were also instituted by the FDA. The FDA's initial approval, of course, gave the drug companies some leverage to delay decisions to withdraw them for further testing. That issue is still pending, and it may well be that the number of deaths or bad outcomes as side effects of the drug use is so low that associating it with the use of the prescribed drug may be coincidental. It is never easy to resolve alleged harm done to a small number of individuals in a large population, and it is similar to the issue of identifying the effects of low doses of radiation on leukemias and other cancers among those subject to local fallout during the Cold War testing of atomic weapons in the American far west.

The thalidomide case also makes an over-the-counter drug more difficult to obtain in some countries. When I had a cold some 20 years ago in Germany, I looked for familiar decongestants I could buy routinely in the U.S. without a prescription in a drug store or supermarket, and I was told by a German clerk that such medications are only available by prescription. Germany now has one of the world's strictest drug regulation programs. I didn't mind the inconvenience and would rather be mildly burdened than have others face the risks of potential harm. I would feel disturbed, however, if I learned of a new drug that was proven effective in animal trials and that was in the initial human clinical trials but was still 3 or more years off from licensing while I or a relative was at risk of being terminally ill with no effective medication for that pathological condition.

The most important lesson is that risk-free medication is virtually nonexistent. If thalidomide had been tested for teratogenicity in rats, it would not have had any harmful effect. It would have had to be tested on rabbits, which, like human embryos, produce deformed limbs when exposed to thalidomide. There may be some future drugs that will slip through present animal tests if a harmful effect is physiologically limited (say to primates) and the usual animal tests all indicate the product was safe. There will also be a tension between the pressures of those who manufacture drugs and those who represent the families of users of these drugs who experienced unexpected side effects unrelated to their illness.

Regulation is an inconvenience, and manufacturers would be happier with much less testing and regulation. But when patient safety is regarded as a higher value, regulation provides the safeguards against self-deception, over-confidence, fraud, indifference, and wishful thinking. It is also worth reflecting on Herbert Keller's remark when he learned that the sedative he worked on with Wilhelm Kunz turned out to be a potent teratogen: "I feel like a bus driver who has run into a group of children."[19]

Notes and References

1. The Insight Team of the *Sunday London Times, Suffer the Children: The Story of Thalidomide* (Viking Press, New York, 1979). This is an impressive study by journalists on all aspects of the thalidomide tragedy. As the date of publication indicates, none of that work could be published under British law until the case was resolved in court (in this case, by settlement).
2. For the latest list of disorders for which prescription of thalidomide is available or pending, see the Food and Drug Administration Web site - www.fda.gov/cder/news/thalinfo/thalidomide.htm. At present, the metabolism of thalidomide is not well understood, and the range of disorders for which thalidomide is recommended (HIV, immune cancers, leprosy, Bechet syndrome, etc.) is very wide and does not have an obvious common metabolic basis.
3. A Web site provided by Scott Gilbert (Swarthmore) provides an updated list of articles on the developmental biology of thalidomide at http://www.devbio.com/article.php?ch=21&id=200.
4. Insight Team, op. cit., p. 16.
5. Ibid., p. 70.
6. Ibid., p. 30.
7. Ibid., p. 14.
8. Ibid., p. 81.
9. Paul M. Wax, "Elixirs, diluents, and the passage of the 1938 Federal Food, Drug and Cosmetic Act." *Ann. Int. Med.* **122** (1995): 456–461.
10. John Lear, "The unfinished story of thalidomide," *Saturday Rev. Literature,* September 1, 1962, pp. 35–40. Lear, who was science editor for the journal, summarizes each of the 50 communications Kelsey and her department received.
11. Insight Team, op. cit., p. 46.
12. Ibid., p. 8.

13. W. Lenz, "Malformations caused by drugs in pregnancy." *Am. J. Dis. Child.* **112** (1963): 99–106; and W.G. McBride, "Thalidomide and congenital abnormalities." *Lancet* **2** (1961): 1358.

14. Insight Team, op. cit., p.126. In Germany, a criminal case was initiated but dropped by the prosecution to permit Chemie Grünenthal to make a settlement (including a substantial sum from the German government).

15. Ibid., p. 15.

16. There are fewer thalidomide victims in the U.S. than in Europe, and hence, they have not been as effective a lobby group in the U.S. as overseas.

17. The FDA has issued strict guidelines for physicians who prescribe thalidomide and patients who use it with a special training session on storage and disposal and strict prohibition of sharing this medication with other people. So far, there have been no reported cases of abuse from users in the U.S. Thalidomide has been prescribed since 1998. The company that makes it, Celgene Corporation in New Jersey, uses the trade name *Thalomid* for its product.

18. Two of my colleagues, one in psychiatry and one in psychology, have told me how difficult it is to get new drugs for treating behavioral disorders because of the thalidomide scare. Despite a recent (2004) scandal over Vioxx (a pain reliever that has led to excess cardiac and stroke deaths), there is a lot of pressure on Congress in the second Bush administration from drug companies and health providers who want more effective drugs for their patients, to relax many of the regulations and threats of lawsuits that drug companies claim scare them off from making that investment. The issue will probably swing between relaxation of regulation when the public memory of disasters grows thin and the reenactment of regulation when a large number of deaths or illnesses occur from some future manufacturer who will rely on its good intentions instead of time-consuming effective testing.

19. Insight Team, op. cit., p. 110.

9

A Synthetic Estrogen with
Harmful Outcomes

IN 1938, THE BRITISH BIOCHEMIST EDWARD CHARLES DODDS and his colleagues at the University of London synthesized an estrogen-like compound and called it diethylstilbestrol (DES).[1] Unlike natural estrogen,
which is a steroid (a compound of four connected rings), DES is not a
steroid and has two separated rings. It also has an additional progesterone-
like activity. It was never patented, and many companies explored potential uses for this hormone. In Great Britain, DES was marketed as a treatment for vaginitis and menstrual disorders.[2] Vaginitis arises when there is
an imbalance among the yeasts, protozoa, and bacteria that normally live
in a balanced state in the vaginal tract. It can lead to irritation, discharge of
mucus, and offensive odors. Dodds had explored the estrogens since 1932,
when he noted that coal tars (especially anthracenes and other polycyclic
compounds) sometimes had estrogen-like activities as well as their better-
known capacity to induce tumors. Dodds was skeptical about the exagger-
ated claims for sex hormones in the 1920s and 1930s, when they were seen
as agents for rejuvenation, longevity, and restored sexual vigor. He worried
that long-time use of these hormones might have unexpected effects. He
died in 1973, shortly after his own DES compound would be associated
with the very conditions he most feared—birth defects and cancers. Ironically, the year before he died he expressed relief, remarking "I must say
that when the thalidomide tragedy occurred I could not help feeling how
lucky we were that there was no hidden toxicity in stilbestrol."[3]

When World War II ended, there was an immense interest in pharma-
cology. The war had seen the production of penicillin as an antibiotic that
surpassed sulfa drugs in efficiency and that had fewer allergic reactions. Bio-

chemistry was also emerging as a relatively new science and finding its way into medical courses, as medical research sought new insights into a variety of diseases. Cortisone was being studied for its effects on arthritis. Many new hormones had been identified for the human reproductive process, and a synthetic estrogen might have value for reproductive conditions.

One theory in vogue at that time was used to explain the relatively high frequency of miscarriages or spontaneous abortions in women.[4] About 20% of women experience at least one miscarriage. It was not uncommon for a woman to experience two or three such unhappy outcomes to a desired pregnancy. The theory sometimes invoked as an explanation for these failures was called "hormonal insufficiency." It seemed reasonable that embryos failed to continue their development if the uterus failed to maintain the right hormonal environment to keep the pregnancy going. Hormonal insufficiency was also then used (before 1957) as an explanation for Down syndrome occurring among older women. On the basis of some observations among women treated with DES for vaginitis and a few poorly controlled studies, women given DES seemed to have fewer abortions. These reports were popularized as promotional stories in *Reader's Digest.* It has been a policy for American physicians to use medications as they see fit unless they are forbidden to do so by law or federal regulation. It was also a policy in the U.S. in the 1950s that if a drug was licensed for one use and considered safe, the manufacturer could promote it for other uses as its successes for other conditions emerged.

By the end of the 20th century, a different explanation of miscarriages emerged. Many of them are chromosomal aneuploidies caused by nondisjunction or polyploidies caused by fertilization of one egg with two or more sperm. Many others are homozygous mutations that abort the embryo because of some essential function during organogenesis—the first 55 days of pregnancy when most organ systems are built from genetic instructions. The majority of spontaneous abortions still have not been explained. No proven hormonal insufficiency is known to be associated with them. In the late 1940s and early 1950s, the chromosomal and genetic contributions to abortion were unknown.

Some physicians who prescribed DES for women who had one or more miscarriages reported reduced numbers of recurrences.[5] No extensive study was carried out, and there were no controls. The quality of medical research in that era did not measure up to the standards of research in the basic sciences. The claims of success were carried in magazine articles and newspaper reports. Some even reported a reduction in premature babies. DES was being hailed as the latest wonder drug following in the footsteps of antibiotics. Some physicians began to prescribe it as a prophylactic medication, to prevent any miscarriages among pregnant women, and this appealed to the women who produced the "baby boomer" generation.[6] Some 200 different companies were selling it in the U.S. as well as in many other countries. About five to ten million women were prescribed DES from 1938 to 1971 whether they had a history of miscarriage or not. Reports began to appear as early as 1953 that DES did not prevent miscarriages or prematurity, but physicians continued to prescribe it.[7]

More disturbing in the early 1970s were reports of a few girls and many more teenage females with clear-cell adenocarcinoma of the vagina. This form of vaginal cancer was previously only seen in middle-aged or older women. It took some time to couple these outbreaks with a common factor shared by their medical history. They were daughters of mothers who had been given DES to prevent miscarriages.[8] The Food and Drug Administration (FDA) issued a stop order in 1971 for the prescription of DES during pregnancy. Because affected daughters did not begin to express their cancers for 10–15 years after the mothers had taken DES, many of their physicians had retired or died, and in some cases, their records were destroyed or lost. In addition, some mothers had died, making it difficult for women born in that era to know whether they were DES daughters. The need to know was important because a DES daughter is at elevated risk for her entire lifetime to vaginal cancer, and she has to have a yearly test to make sure the cancer is not present. The risk is not trivial. There is a 40 times greater likelihood of such daughters to develop clear-cell adenocarcinoma than daughters whose mothers did not receive DES as a prophylactic.[9]

A second feature of the DES bad outcome involved the impairment of the uterus of many of these daughters. They had a higher than normal risk of a malformed uterus. This included deformed shape, septation of the uterus, or displaced location. It contributed to a higher incidence of infertility of these women and added complications to their deliveries if they did become pregnant.

For all three of those involved—the physicians who prescribed the DES, the mothers who used it, and the daughters at risk for cancer and reproductive problems—this created a psychological depression.[10] No physician likes to see patients harmed by a medical procedure that turns out to be erroneous. They feel as if they have violated their oath to do no harm. Mothers feel guilty that they put their daughters at risk for health problems they would not have had. The daughters feel depressed that their lives will be shadowed by the risk of vaginal cancer.

There was no criminal or unsavory behavior by the medical profession in endorsing DES for their patients. Nor did the drug companies have any doubts about their optimism that they truly had a wonder drug like penicillin to offer the public. At the time DES was licensed in the U.S., it had been tested and used in Great Britain for vaginitis and menstrual difficulties. This was the pre-thalidomide era, and there was no prior history of teratogenic effects of prescription drugs. If there is blame to identify, it is in the unregulated permission of physicians to prescribe a drug for whatever they desire once a drug is licensed. A second failure was the lack of long-term controlled experiments in testing the drug for efficacy. Although legally the drug companies were not obliged to do such experiments then, ethically it is difficult to imagine the companies approving use of the drug for miscarriage prevention based essentially on anecdotal evidence. A third failing is the recurrent belief that science will prevent aging, confer immortality (or greatly extend individual life expectancy), and maintain sexual desire and competence well into old age. These are unproven fantasies associated with El Dorado, Shangri la, Caucasian yoghurt eaters, and even earlier quests like that undertaken by Ponce de Leon for a fountain of youth in the New World. It suggests some combination of self-deception, ignorance, or wishful thinking by those physicians whose optimistic reports stimulated the drug companies. In many respects, they showed the same naiveté as Dr. Harry Clay Sharp showed when he believed vasectomies were a good treatment for the prevention of masturbation and degeneracy.

Notes and References

1. E.C. Dodds, L. Goldberg, W. Lawson, and R. Robinson, "Oestrogenic activity of alkylated stilboestrols." *Nature* **142** (1938): 34.
2. R.J. Apfel and S.M. Fisher, *To Do No Harm: DES and the Dilemma of Modern Medicine* (Yale University Press, New Haven, 1984), p.14.
3. F. Dickens, "Edward Charles Dodds (13 October 1899–16 December 1973)" *Biographical Memoirs of the Fellows of the Royal Society* **21** (1974): 227–267. Dodds (who used Charles as his given name) had a distinguished career as a medical biochemist and studied tumor-inducing and tumor-inhibiting compounds. The quote is from p. 247.
4. The cause of miscarriages is still unknown. Very early ones have a high incidence of chromosomal abnormalities, including trisomies and polyploidy from

polyspermy. Other causes are suspected but not proven, including infections, immunological incompatibility, or hormonal disturbance. It was this latter unproven concept that motivated drug companies to promote DES as supplementing an alleged insufficiency. In 2005, some physicians have revived (without evidence) hormonal insufficiency as a cause of miscarriages.

5. Apfel and Fisher, op. cit., pp. 20–23. The authors cite the influence of several papers from Harvard Medical School produced by George Smith (Ob-Gyn) and his wife Olive Watkins Smith (Biochemistry) in the 1940s and early 1950s. The Smiths claimed DES had positive outcomes for pre-eclampsia, progesterone increases, pituitary stimulation, and prevention of miscarriages.

6. Ibid., pp. 38–39.

7. W.J. Dieckmann, E.M. Davies, L.M. Rynkiewicz, and R.E. Pottinger, "Does the administration of diethylstilbestrol during pregnancy have therapeutic value?" *Am. J. Obstet. Gynecol.* **66** (1953): 1062–1081. Dieckmann studied over 2000 women given DES or placebos for first-time pregnancies. The study involved double-blind control methods. They found no benefit from the use of DES.

8. P. Greenwald, J.J. Barlow, P.C. Nasca, and W.S. Burnett, "Vaginal cancer after maternal treatment with synthetic estrogen." *N. Engl. J. Med.* **285** (1971): 300-392. Also see A.L. Herbst, P. Cole, T. Cotton, S. Robboy, and R.E. Scully, "Age-incidence and risk of diethylstilbestrol-related clear cell adenocarcinoma of the vagina and cervix." *Am. J. Obstet. Gynecol.* **128** (1977): 43–50.

9. E.E. Hatch, J.R. Palmer, L. Titus-Ernstoff, K.I. Noller, R.H. Kauffman, et al., "Cancer risk in women exposed to diethylstilbestrol in utero." *J. Am. Med. Assoc.* **280** (1998): 630–634.

10. Apfel and Fisher, op. cit., Chapter 4 "A quiet trauma," pp. 59–68.

Part 5

THE NECESSITY OF
REGULATION TO PROTECT
THE ENVIRONMENT

THE POST-WORLD WAR II ENVIRONMENTAL MOVEMENT WAS, in large measure, launched by the writings of Rachel Carson, especially her classic book, *Silent Spring*. Carson wanted government regulation and self-regulation by scientists. She was not an extremist who sought a back-to-nature movement, nor did she advocate the dismantling of our technology-dependent society. She wanted people to be aware of abuses and neglect, and in this effort, her work was successful. Any new movement spawns a spectrum of participants, some of whom we would label extremists. Environmentalists who destroy laboratories or burn crops that are being tested for their characteristics (such as genetically modified foods) are examples of such extremism that Carson would have repudiated. Unfortunately, many critics of the environmental movement judge it by its extremists.

In Carson's analysis, the new, post-World War II use of pesticides (she called them biocides because they were not targeted to just one pest) and herbicides led to environmental damage and health concerns for the public. Her solution was more regulation, a policy based on reason, and an informed team of scientists who would include those aware of the ecological and evolutionary consequences of widespread use of these agents. Understandably, there will be a tension between advocates of public safety and environmental protection and advocates of business free of the costs, time, and paperwork that regulation involves.

A similar tension exists in the new field of genetically modified foods, the second theme explored in this section. Whereas farmers have long modified foods by selection and breeding, the term genetically modified foods since the 1990s refers to the direct insertion of genes (natural or synthetic) from one species into another species. Unlike Carson's concerns for a damage that she could document, critics of genetically modified foods are more fearful of the rare or novel chimerism of organisms with genes from other species. They have little to go on for evidence of harm because the field is new, but they argue that prevention is a wiser policy than imposing regulations after some damage to health or the environment occurs. This attitude changes the debate from actual evidence based on historical events to potential damage based on difficult-to-test suppositions which may reflect unwarranted fears. In such debates, it is not uncommon for both sides to be ignorant of what can happen, but the advocates tend to be harsher in their assessment of their foes than they are in the assessment of their own case.

In all likelihood, the debate on genetically modified foods will eventually disappear, just as the debates over recombinant DNA technology in the pharmaceutical industry in the 1970s and 1980s have disappeared because of a long history of regulation and safety. I did not include an extended discussion of the recombinant DNA controversy because that has been well explored and resolved. Those interested in how this controversy arose from scientists's own fears and how most of those fears turned out to be both unfounded and carefully monitored and regulated will benefit from reading J.D. Watson and John Tooze's source book.[1]

Notes and References

1. J.D.Watson, and J. Tooze, eds. *The DNA story: A documentary history of gene cloning* (W.H. Freeman, San Francisco, 1981).

10

Pesticides and the Environmental Movement

THERE ARE TWO INFLUENTIAL NONSCIENTIFIC OR RELIGIOUS WAYS that humanity has perceived for the relationship between people and the environment. The oldest is based on a biblical interpretation that we have the mandate (i.e., "dominion...over all the earth," in the King James translation of *Genesis*) to use nature as we see fit.[1] Nature in this view is a preserve for us to live in, to exploit, to respect, or to dominate, depending on our interpretation of nature. It was part of the trade-off after the expulsion from Eden. For those who saw nature as threatening, wild, and acting as an obstruction, nature had to be tamed. For those who saw it as convertible to human benefit, it was ours as a resource to use. Opposed to this is a second view, often championed by Native Americans in their religious traditions. It is a relationship between humans and their environment that could be called a stewardship. Both humans and the living species in the communities around us provide a mutual benefit if we respect nature. For Native Americans, the world of plants and animals is highly spiritual and anthropomorphic. For most of those who were raised in the biblical tradition, that spirituality of nature is absent.[2]

There are also two secular ways we have interpreted nature. It may well be that these past religious traditions have played a part in shaping both of these perceptions. The first interprets nature as ours to use. We can cut down its wilderness to make way for farms. We can dam its rivers to regulate the supply of water to our towns and farms. We can dig into the earth and refine its ores and burn its fuels. We can fish the waters, shoot the birds, trap the small animals, and domesticate the animals and plants useful to us. We can cut deeply into forests and use their lumber to construct our homes. Opposed to this is a view that began in the mid-19th century with the efforts of Henry David Thoreau, followed by John Muir and Theodore Roosevelt. They founded the conservation movement and recognized that parts of nature should be shielded from commercial use so we can enjoy its natural beauty and allow species to live that would otherwise become rare or extinct. This stewardship had an aesthetic or romantic attraction for many scientists involved in the conservation movement; some elevated nature to the status of a church. For the greater part of the 20th century, the secular view of conservation has prevailed.[3]

These views are not mutually exclusive. Those who want to profit from using nature, such as loggers and others involved in the forestry industry, recognize that too zealous a destruction of forests will drive them out of business. Overzealous fishing or hunting will lead to species extinction. Not all people are wise and thoughtful, however, and society has had to put regulatory restraints on those who erroneously believe that nature is restorative, that nothing goes extinct, and that everything will be replaced in due time once the land has ceased to yield a profit or ceased to provide a wholesome environment. At the same time, conservationists have recognized that we cannot go back to a hunter–gatherer state of existence. We have our civilizations, and there must be some appropriate use of the natural world to allow humans to enjoy those benefits. In both cases, the degree to which nature can be used and how much regulation there should be are controversial.

Pre-World War II Developments

Until the end of World War II, the pace of environmental change was relatively slow. Most of it involved a gradual encroachment by chopping down trees and clearing land of its tangled underbrush and hauling away rocks to make way for farmland. The American wilderness that Audubon first observed in the 1830s was largely gone by the 1930s.[4] The damage done was chiefly habitat removal by human replacement, mostly farmers. The passenger pigeon may have disappeared from its East Coast grounds, but in our practical utilitarian ethics, we more than benefited by having a cultivated, fertile countryside producing generous amounts of food.

The pace changed as the chemical industry was put to use during World War II and introduced a number of organic compounds that served as insecticides and herbicides. For the insecticides, there was an immediate human health consideration. Delousing our own troops and those inhabitants in whose countries our armies marched was very effective. Lice and fleas carried microbes that caused typhus. Mosquitoes carried the protozoan microbes that caused malaria, and in both the Mediterranean and Pacific areas where our troops were sent, malaria was abundant. Paul Müller, a Swiss chemist at Geigy Pharmaceuticals, recognized in 1939 that DDT (diethyldichlorophenyltrichloroethane; known to chemists by its formal structural name, 1,1,1-trichloro-2,2-*bis*(*p*-chlorophenyl)ethane), which had first been synthesized in 1873 by an Austrian chemist, was a potent killer of insects. Müller won the 1949 Nobel Prize in Physiology or Medicine "for his discovery of the high efficiency of DDT as a contact poison against several arthropods." Two of the more popular trade names for DDT when it was marketed in 1942 were Gerasol and Neocide.

When the war ended, chemical companies realized they had a huge potential market for pesticides. Insecticides could be used in the home for pests like cockroaches and bedbugs,[5] in the yard to protect roses from the voracious appetites of Japanese beetles, and on crops to kill corn borers and other insects that could damage 5–10% of a crop. Pesticides could be sprayed on ranches to eliminate weeds competing with grass, and this would reduce the cost of raising cattle. By 1960, chemical industries in the U.S. were manufacturing an annual output of about 638 million pounds of insecticides.[6]

Of significance for this new habit of using chemicals on a large scale to destroy pests of various kinds were the following factors:

- Organisms do not evolve independently of other organisms. Each organism lives in an ecosystem. The destruction of a category of plant or animal life will likely have ecological consequences for other species living there.

- Organic pesticides are frequently fat-soluble, and they may be concentrated or stored in larger amounts in fatty tissues than in the soil, the water supply, or other environmental locations.

- Some organic pesticides may persist in the soil (or in an animal body) for many years and thus continue to enter the food chain long after a spraying program has been discontinued. The stored pesticide in an organism may result in chronic health problems.

- The manufacture of organic pesticides or their dispersal by airplane or by pumped aerosols may expose workers and those who are hired to apply it to high doses of the pesticides. These can be toxic and sometimes fatal. They can also lead to chronic illness.

- A pesticide intended for a particular pest may also kill beneficial insects that other farmers may depend on (such as bees), or they could lead to starvation or illness of birds that feed on the insects.

- Some pesticides may have unexpected biological activity such as inducing mutations, inducing cancers, or interfering with human or animal metabolism.

- Organisms that are exposed to pesticides vary in their genotypes, and a few may survive, leading to resistant forms to any of the pesticides used. There are always new mutations arising in the reproductive cells of a population, and this will also permit the lucky few to survive. It will force the user of pesticides to buy different pesticides or use higher doses.

Very often those who synthesize a compound that acts as a pesticide may not be aware of the biology taking place when the product is used. The same is true for those who do the marketing and who see things from a commercial perspective. If manufacturers do not hire scientists or consultants who have that knowledge, they may act in ignorance. Their intentions are usually good. They see a product that will eliminate a pest. They think of this as beneficial to humanity. What they do not see is the connected way in which evolution and nature work. Bad outcomes of this sort are usually prevented by regulation. However, that regulation is not likely to be enacted or required unless bad outcomes have occurred.

Rachel Carson Shifts Conservation to Environmentalism

The person who changed the conservation movement into the environmental movement was Rachel Carson (1907–1964).[7] She grew up in a stressed, crowded household supported by her father, an unsuccessful businessman, who died unexpectedly; this left her with a family to support. She was also a very talented student and, while her father was alive, tried to manage on scholarships (very rare then in the 1920s) as she completed her B.A. at the Pennsylvania College for Women. She had started out as an English major but so enjoyed her biology course that she switched to graduate with a biology major. She had hoped to get a Ph.D. at Johns Hopkins University but only had the funds and the time to complete a Master's degree. She withdrew in 1934 in the depths of the Depression when money was scarce and her family was desperate for her support. Beginning in 1936, she worked as a publicist for the Fish and Wildlife Service writing brochures and did as much field work and laboratory work as she could. Her writing was so good that it began to attract notice as she supplied magazines with articles about the natural world. In 1951, she obtained a Guggenheim fellowship and wrote her first best-seller, *The Sea Around Us*. She also became aware of the changing ways people were altering the environment. She began collecting reports from state committees, wildlife management groups, and Farm Bureau associations; and newspaper clippings on accidental deaths involving the use of pesticides, sicknesses attributed to the manufacture or exposure to pesticides, and most of all, the profound ways in which rivers, lakes, and the landscape, as well as the life they contained, were being altered by the millions of pounds of pesticides entering the ecosystem. In 1962, while ill from arthritis and breast cancer, she completed *Silent Spring*, her most famous book and no doubt one of the most influential books on science and society ever written.[8]

Carson wrote her book as a sustained argument against the chemical industry and those who used insecticides, fungicides, and herbicides without knowledge of what they were doing to organisms or the habitats in which they lived, or even to the people who used these products in innocence. She argued that evolution is generally a slow process, and organisms adjust to their environmental changes over long periods of time. For that reason, any ecosystem is in a balanced state with a mutualism of activities of a myriad of organisms. She describes in detail how the soil has microscopic bacteria, fungi, and algae, as well as nematodes, mites, insects, and earthworms, that keep it aerated and supplied with the degraded litter of leaves, twigs, dead grass, and dead animal life. When we walk on soil we are walking on a carpet of life, dead and alive. It takes decades or centuries to grow a few inches of soil.[9] She also describes in detail how all water is connected and recycles, with the groundwater on a sprayed acreage carrying the pesticides to locations miles away as the water relentlessly enters wells, streams, and lakes.[10]

When people spray pesticides, they do not think about the bad outcomes of their spraying. To the sprayer, the pesticide is only being used on a backyard garden, or only on a golf course, or only on the 100 or more acres of a cornfield. But in Carson's analysis, they are unwittingly spreading the pesticide through the groundwater, through mists, through the food chain, and even through the bodies of their own and their neighbor's children. Carson was more than a polemicist. She marshaled her evidence from her clippings and the testimonies reported in hundreds of sources. Her indictment was severe:

- Organic chemicals should not be described as insecticides but as "bio-cides" because their killing is not selective against a particular pest (e.g., a corn borer) but affects virtually all the insects in the field, including honeybees and butterflies beneficial to the farmer.[11]

- Organic chemicals for commercial use are new to nature, and the long-range effects on human health or the environment have not been tested by the chemists who manufactured them. The living world has not had time to adapt to them.

- The public is being misled, and the manufacturers may be self-deceived, in believing that a single spraying will solve their problem. Because of genetic variation, new mutations, and resistance to pesticides among a small number of insects, fungi, bacteria, or other targeted pests, there will be a return of resistant populations of the pest, making future use of the pesticide ineffective.

- The natural environment is being turned into an artificial environment because users apply indiscriminate spraying instead of selective spraying. When roadsides are sprayed to eliminate a particular weed as a pest, the wildflowers are also destroyed, and the roadsides are replanted with grass or left with withered blight. Planting with grass is more expensive because it requires that the roadsides be mowed regularly.

- Organic pesticides have metabolic consequences as they move through the food chain, concentrate in lipid-rich tissue, and may end up many hundred times or even a thousand times the concentration that was sprayed. The indiscriminate and indirect killing of insects unrelated to the intended target can lead to starvation of those organisms depending on them or poisoning of those organisms that eat them.

Carson described the dilemma of living in an age when intelligent people can do their work with good intentions and be unaware of their work leading to bad outcomes: "This is an era of specialists, each of whom sees his own problem and is unaware of or intolerant of the larger frame into which it fits."[12] That ignorance, she warned, has led to "the pollution of the total environment of mankind."[13] She included radioactive fallout, the dumping of industrial and municipal wastes into rivers and oceans, and the runoffs from sprayed ranges and farms as evidence for this global assault on the environment (much of it unintended).

Best known in her book is the chapter that opens it, "A Fable for Tomorrow," which gave the title to *Silent Spring*.[14] She describes the grad-ual disappearance from farms of songbirds that feed on insects who carry a load of amplified DDT and other pesticides. She describes the disappearance of birds eating fish that eat smaller fish that eat plankton with concentrated amounts of pesticides. The cause of their illness and disappearance was still debated when Carson wrote her book. Her surmise that eggshell thinning was the major factor in preventing a new generation of birds from being born has been proven in the decades since her book appeared. The Environmental Defense Fund took up this issue in the courts and succeeded in getting the publicity and court decisions to force a ban on DDT.[15]

Complicating the DDT controversy was Carson's claim that DDT causes cancer, which scared a lot of people. There was no direct proof then that this was so, but the Environmental Protection Agency presently lists DDT as a probable carcinogen. Carson was astute in singling out the American bald eagle, our national symbol, as a raptor at the top of its food chain. She argued that its plummeting numbers, making it an endangered species, were associated with elevated DDT concentrations. The peak use of DDT in the U.S., in 1959, reached 80 million pounds, mainly on agricultural crops.[16]

Carson made a plea for the regulation of the chemical industry. Chemists are just not aware of the damage their products do to other living things. Manufacturers have a faulty understanding of how much arsenic, heavy metals, and organic chemicals can be dumped into the rivers and soil of those places where these products are made.

The reaction to Carson's *Silent Spring* was a firestorm of criticism and praise. She was seen by the chemical industry as a hysterical woman. Book reviews interpreted her work as "angry and shrill."[17] Former Agricultural Secretary Ezra Taft Benson accused Carson of being a Communist (this was the Cold War era), echoing a charge of a chemical industry representative who accused her of supporting a Communist plot to reduce American and European agriculture to parity with the Communist world.[18] It was ironic to Carson that applications of insecticides intended to increase crop yield were being carried out while the government was providing huge subsidies to farmers not to plant their crops because bumper crops would drive down the price of American wheat, corn, and soybeans on the domestic and world market.

Fortunately for Carson and the American public, Carson was vindicated by hearings by the President's Science Advisory Council which confirmed the careless way in which pesticides were being used.[19] John F. Kennedy read her book and was moved by it. He was instrumental in getting these hearings. Their 1963 report "The Uses of Pesticides" called for "the elimination of the use of persistent toxic pesticides" as well as federal research regarding these chemicals. The Committee noted that "until the publication of *Silent Spring*, people were generally unaware of the toxicity of pesticides...."[20] This report led to the formation and coalition of environmental groups, including the Environmental Defense Fund (established in 1967), the Legal Defense Fund of the Sierra Club, the National Wildlife Federation, and the National Audubon Society. Their court actions sought relief from the overuse of pesticides. The combination of ecological damage and human health risks proved too powerful a concern by voters for legislators to ignore. By 1970, Congress established a U.S. Environmental Protection Agency (EPA). In 1972, the use of DDT, the most notorious of the agents doing ecological damage, was banned by the EPA.

Carson's legacy lasted far beyond her death in 1964. Her work paved the way for environmentalists to sound the alarm over the chemical wastes produced by Hooker Chemical that were dumped locally in landfills which were later paved over for developers to build housing in Love Canal near Niagara Falls in New York state. The wastes included 200 tons of dioxin and 200 other chemicals. Although Hooker notified the city in 1953 about the contents of the landfill, the city built a school and housing development on the site when the land was signed over to them. A high frequency of birth defects and chronic illnesses made the residents aware that they were living in a toxic environment. In 1980, the community was condemned and evacuated by federal emergency authority. More legislation in the form of the Superfund program of the EPA was enacted in the Comprehensive Environmental Response, Compensation, and Liability Act of December 11, 1980, to deal with situations like Love Canal.[21]

Environmentalists today have tried to raise public interest concerning the loosening of federal standards for pollution since 2000. In general, those with an economic set of utilitarian values point out the importance of jobs and competitive advantage against foreign industries. No industry wants to spend significant amounts of money on such environmental measures if its products are not immediately hazardous to its workers or to the community. Meeting environmental standards leads to higher prices, and these industries fear foreign producers will drive them out of business. Environmentalists argue that the costs to society in paying for the illnesses of workers and residents, as well as the hardships imposed by death of those who support families, place a greater burden of financial lost. Property values in contaminated communities, especially those declared as Superfund sites, plunge in value and can turn the communities into ghost towns. These hidden costs of pollution end up being paid by the victims or by the taxpayers who are subsidizing companies with indifferent attitudes to environmental health. Environmentalists also remind us that it is not just money that is at the center of the debate, it is health for humans and a place for living things to grow and flourish that must be taken into account.

I believe that it is not an either/or debate. Both industry and the conservation movement can work out ways to coexist through the effective use of government regulatory agencies. These agencies work best if they are not staffed politically but are staffed through a merit-based program like Civil Service. A politically staffed agency is likely to shift policy whenever an election takes place. Experts hired by competitive examination for fixed or lifetime appointments are more likely to represent the public (for whom they are asked to serve) than the political administration in power at the time.[22]

There are now dozens of organizations serving as educational groups, lobbying groups, and watchdogs for environmental causes. As in any mass movement, some will be less well informed than others, and their motivations will vary for protecting human health and the environment. In a democracy, that is the way it has to be and we must keep as informed as we can be listening to a spectrum of environmental voices that include some who downplay environmentalism as a romantic movement lacking scientific evidence of harm and some who see everything applied to the environment other than a return to nature and organic farming as a crime against nature and humanity.

Notes and References

1. *Genesis* 1:28. The word dominion implies "absolute authority and control over," as one biblical commentator notes (*Life Application Study Bible King James Version*, see p. 7. [Tyndale House Publishers, Wheaton, Illinois, 1997]).

2. The concept of dominion over the earth was taken literally until the environmental movement made most of humanity (including scriptural literalists who adopted the idea of stewardship) aware that some care had to be paid to the environment. Occasional regulation was made by the church by overzealous use of the term dominion. Thus, when Native Americans were first discovered by Hispanic conquerors, some conquistadors felt that they could kill them at will or put them into slavery without baptizing them because they were considered to be wild animals over which they had dominion. The church in the early 1500s issued a Papal Bull affirming that the Native Americans were fellow human beings, not wild animals, and that they had to be baptized if enslaved and could not be killed without prop-

er cause. There is also some inconsistency on the part of the earliest inhabitants of the New World in their relation to the land. Far from being a stewardship, it involved, according to many archaeologists, a slaughter of the megafauna (e.g., giant sloths, mastodons) that existed after the recession of the last Ice Age. Some Evangelical Christians have recently adopted the Native American view of stewardship and have surprised conservative business interests and their allies in Congress by siding with the environmentalist movement.

3. For an account of John Muir's life and his role in the conservation movement, see John Muir, *The Wilderness Journeys* (1913) (Paperback reprint, Canongate Books, Edinburgh, 1986).

4. William Souder, *Under a Wild Sky: John James Audubon and the Making of the Birds of America* (North Point Press [a division of Farrar, Straus, Giroux], New York, 2004). This superb biography of Audubon also gives a remarkable portrait of the American wilderness in Audubon's youth.

5. It was remarkably effective against bedbugs when I was a boy and when my parents used it in the mid-1940s. Prior to that, there would be bimonthly scrubbings of the bed springs with kerosene, which wasn't very effective. Bedbugs were totally eliminated after one application of DDT. That was not the case for roaches, which somehow had better genetic diversity, enabling resistant roaches to make a comeback within a year of its first application.

6. Arlene Quartiello, *Rachel Carson: A Biography*, p.84 (Greenwood Press, Westport, Connecticut, 2004).

7. Quartiello's biography is intended for high school students, but it has an excellent overview of her life and is done in a scholarly style with references.

8. Rachel Carson, *Silent Spring* (Houghton Mifflin Company, Boston, 1962).

9. This was first recognized by Charles Darwin, who demonstrated that slow growth experimentally using earthworms in his estate at Down, England. His work on soil is in Charles Darwin, *The Formation of Vegetable Mould, through the Action of Worms, with Observations on their Habits* (John Murray, London, 1882).

10. In her words, "all the running water of the earth's surface was at one time groundwater." Carson, op.cit., p. 42.

11. Ibid., p. 8.

12. Ibid., p. 12.

13. Ibid., p. 39.

14. Ibid., pp. 103–128.

15. Charles Wurster, "The power of an idea" and "The last word." In *Acorn Days: The Environmental Defense Fund* (M.L. Rogers), pp. 44–53 and pp. 178–183. (Privately printed, New York, 1990.)

16. U.S. Environmental Protection Agency, "DDT ban takes effect." Press release dated December 31, 1972. http://www.epa.gov/history/topics/ddt/01.htm.

17. Quartiello, op. cit., p. 108.

18. Ibid., p. 107.

19. Ibid., p. 111.

20. Presidential Science Advisory Committee, *The Uses of Pesticides: A Report of the President's Science Advisory Committee* (GPO, Washington, D.C., 1963).

21. More information about the Love Canal can be found in the Love Canal Collec-

tion, University Archives, University Libraries, State University of New York at Buffalo, http://ulib.buffalo.edu/libraries/projects/lovecanal/.

22. There is a huge literature of periodical articles and books on the environmental movement. A significant portion of biologists work in fields tied to that movement, and some are active in promoting those views in their courses and in their communities. The issues include the loss of biological diversity through species extinction; the actions of specific chemicals on fish, invertebrates, and other life forms that live in rivers; the molecular tools needed to identify how a pollutant leads to tissue or organ damage; and the way in which carbon cycles, nitrogen cycles, and other essential parts of the earth's renewal can be altered (often for the worse) by neglect or ignorance (global warming is just one such major issue associated with carbon dioxide emissions from industry).

11

Genetically Modified Foods— As Usual

WHEN I TAUGHT FOR TWO WEEKS IN A SKIING VILLAGE, Bakuriani, in the Republic of Georgia in the Caucasus, I noted how puny and irregular the apples were, and when I tried eating one the core was rotted black with a fungus. The grapes were about the size of my smallest fingernail and with pits, but there were some good foods, too. One was buckwheat used as a side dish during dinner, although it was small comfort when the beef had the toughness of shoe leather. I thought about the American supermarkets where I shop for food and wondered if I put the deformed, small, and questionable apples in a bin next to some freshly picked Macintosh apples all gleaming and relatively symmetrical, which would people buy? And if those miserable discolored grapes were next to seedless large green or red grapes, which would people prefer? We may feel uncomfortable that competitive businesses narrow our choices (but they sometimes increase them vastly) and that agribusiness rather than a family farm is providing the bulk of our food, but there is little doubt in my mind that consumers prefer foods they can trust not to be rotten or aesthetically repulsive.

Farmers have always wanted to make a living by producing more food than their families can eat so they can sell their surplus and buy other things. Over the centuries, they have selected the seeds of better varieties for their next year's crop, and they have mated up the more robust animals of their farms, eating the less hardy ones before they could breed. This process of selection over millennia has produced such unique varieties that many of our domesticated plants and animals have no living wild ancestors. There is no wild maize—there may be teosinte, but it is so tiny compared to the maize grown through hybridization that we would not recog-

129

nize it as an ancestor and throughout the 20th century botanists debated whether it truly was the source of modern maize. We rarely think of the origins of our foods. Navel oranges and many seedless fruits owe their origin to mutations that arose and were vegetatively propagated by grafting.

In the 1920s, the Russian botanist, N. I. Vavilov, demonstrated that the centers of origin of domesticated plants had the most genetic variation. That made sense, because as farmers took their best stock of seeds and animals and moved on to new territories, they took a selected strain they desired and from this a proliferation arose with less variation than was present in the place that these commodities first grew. Loss and gain of variation is as old as the history of agriculture. Many of our modern varieties were developed by selectionists with a good idea for people's tastes and prejudices for food. Luther Burbank did not hesitate to chop down and replace acres of trees if he found one tree with better fruit on it. Dozens of varieties of fruits and vegetables we take for granted were selected by his patience and enthusiasm. Even more have been produced through a knowledge of Mendelian genetics, where genes have been combined to produce almost anything that sells on the market—cattle with milk production that is staggering; cattle with marbled fat in their steaks for those cholesterol-indifferent gourmands who love a good steak; turkeys that can't fly; chickens that lay immense numbers of eggs; pork that is lean; fungus-resistant strains of cereal grains; and rice strains that can grow in brackish water.

Farming is practical. From an evolutionary perspective, almost all the food we purchase, if let loose in nature, would be losers in Darwinian natural selection. These domesticated plants and animals have been bred for the market, not for survival. This is true whether the varieties are called organic or whether they are mass-produced on multi-thousand-acre farms.

Mendelian Genetics Dominates Modern Agriculture

In the U.S., a system of agricultural colleges and field stations was established by law beginning with the Land Grant College Act in Abraham Lincoln's administration. When Mendelism was rediscovered and extended to many plants in 1900 by continental botanists, American scientists at these colleges quickly began to study maize, beans, tomatoes, potatoes, tobacco, poultry, and livestock for commercially useful traits.[1] In the process, they made many important contributions to basic science, as well as some significant contributions to agriculture. At Cold Spring Harbor in New York,

George Harrison Shull in 1908 did an analysis of a field of maize and showed how varied the genetics was. He then inbred two varieties of maize and crossed them to one another. To his surprise, the hybrid plant that arose was immensely more robust in size, rows of grains, and length.[2] This phenomenon later became known to geneticists as hybrid vigor and heterosis. Shull saw it, initially, from a practical perspective, and he proposed hybridization as a way of producing seeds that could be sold to farmers who would benefit by hugely increased yields on their farms. That advice was tested by Edward Murray East, who used four such strains and crossed the two hybrids from each pair to produce an even more remarkably useful hybrid corn.[3] By the 1930s, hybrid corn was widely sold throughout the U.S., and it made American farmers the food suppliers of the world.

Soon plant geneticists at agricultural colleges were producing better varieties of fruits, vegetables, and livestock. They had more flavor, texture, appearance, and other attributes that customers preferred. Most of these successes were done by controlled breeding studies. After 1927, when Muller and Stadler first introduced the induction of mutations by X rays in animals and plants, breeding stations in Europe were inducing mutations they selected for improved cereal crops.[4] In 1944, Milislav Demerec at Cold Spring Harbor extended Muller's radiation technique to the mold *Penicillium* and isolated artificially induced mutations that greatly increased the yield of its antibiotic, which dramatically led to mass production of that antibiotic for a variety of bacterial infectious diseases.[5]

What these efforts used was a system of vertical transfers of genes from a parent to an offspring by breeding analysis. Such changes require a knowledge of Mendelism and a geneticist's skill in isolating or combining different genetic traits. In addition to selecting traits by vertical transfer of genes, some plant geneticists used chemical means to change the chromosome number of plant cells, as in the technique developed by A. F. Blakeslee and O. J. Eigsti in the 1930s for inducing polyploidy using colchicine.[6] Triploid watermelons are seedless, and some seedless varieties of fruits can be produced by this method. The seed to generate them has to be produced anew each generation.

Note that both artificial and natural means can be and have been used by scientists to produce new varieties; this has been an on-going activity throughout the 20th century. The organic farmer uses virtually the same synthetic or domesticated plants and animals as the commercial farmers who douse their soils with anhydrous ammonia and spray their plants with pesticides.

Horizontal Transfer of Genes Is Accelerated by
Molecular Techniques

Occasionally, genes are passed from one species to another by a vector. It might be an aphid that deposits bits of chewed tissue from one plant into another plant's stem. It may be a parasite that lives within a cell and receives or donates genes from or to its host.[7] Natural transfer of genes is relatively rare, but life on earth has been several billions of years in the making and there have been immense transfers of such genes across the phyla over the eons. Whatever the beneficial, or more likely detrimental, effect such transfers have on individual plants, few have been observed in recorded times.

In the 1980s, techniques for transferring genes from one cell to another were introduced by scientists. Originally applied to bacteria and yeast, they have been extended so that any species can harbor genes of another species.[8] The tools for doing this involve recombinant DNA, bits of DNA with appropriate tags that allow them to be picked up by a host cell and incorporated. They can then be isolated and developed so they become part of the reproductive tissue and a line of vertical descent that can transmit that gene to future generations. By such means, cattle can be created that produce pharmaceutical products such as human clotting factors for those with hemophilia. They can produce vaccines for immunization against many infectious diseases. They can produce human hormones that once had to be extracted from human cadavers.

Most of the effort of these studies is going into agriculture. In many parts of the developing world, there is a great loss of food from fungal infection, bacterial rotting, and insect devastation. By inserting genes, making the plants resistant to these various destructive agents, molecular agrobiologists hope to make a contribution to the survival of people in these parts of the world. They are also applying the same techniques to prevent crop loss in the U.S. by introducing natural genes from one resistant species into an unprotected species.

Resistance to this new technology involves both legitimate fears and a somewhat confused picture of what is natural.[9] The legitimate fears include the following:

• The product produced by the inserted gene may act as an allergen to some people. They might be unaware that the food product, formerly considered safe, has been modified.

- The product produced by an inserted gene for a commercial plant may confer resistance to an herbicide that destroys competing weeds. If this gene enters the weeds by horizontal transfer through a natural vector, it could lead to resistant weeds which could spread elsewhere.

- The product produced could be toxic to unintended species such as butterflies or bees that use the pollen of these plants.

These are legitimate concerns and can best be solved by testing the altered plants in laboratories and later retesting them in isolated or contained fields to see whether there is a risk.

Some critics of genetically modified foods are not so much concerned about harmful effects of the altered plants themselves, but of the impact they will have on the practice of agriculture. The modified seeds are being made by large industrial companies (Monsanto is one of the targets of these critics), and small farmers fear they will be driven out by the larger farms' use of these seeds. They fear that seed varieties will diminish as one profitable line is scaled up to sell to most farmers in a country or in a particular climate. Loss of variation is a constant worry of critics of agriculture. In the Soviet Union in the 1920s and 1930s, N. I. Vavilov tried to solve this problem by storing varieties of seeds from around the world, and his Leningrad laboratory kept them going even during the war years when its citizens were starving. If farmers do not plant such varieties, they can quickly become lost. The same issue of commercial seeds was raised when hybrid corn and other laboratory-developed strains in mid-20th century were widely used around the world. Many Third World farmers thought it necessary to abandon their favored local varieties to plant cash crops that wealthy Americans and Europeans desired.

A major difference between the concerns of the 20th century and those related to genetically modified food in the 21st century is the potential for companies to design seed that is sterile. This forces a farmer to buy the commercial seed for life. This, too, is not new. Hybrid corn has to be made anew each year, and it has virtually replaced strains of corn in use in the 19th century. Farmers were quite happy with hybrid corn because it vastly increased their incomes, and the initial fears evaporated. Some people, who resonate with a romantic view of nature, find the production of sterile seeds abhorrent, but Eigsti and others in the 1940s began to produce sterile seed with colchicine to produce seedless watermelon. For the user, it may not matter whether seedless grapes from Mexico or Chile were prepared by colchicine, arose by genetic accident, or were designed by Monsanto using recombinant DNA technology. Certainly they would not differ by appearance or taste.

Quite different from the commercial concerns is the concern about genetically modified foods being unnatural and therefore inherently untrustworthy or dangerous. It is difficult to understand this fear of the unnatural. We wear eyeglasses (unnatural), have pacemakers (unnatural), supplement our dead thyroid with synthetic thyroxin (unnatural), wear hearing aids (unnatural), and eat immense amounts of unnatural foods. Who eats teosinte? Who would spurn a sterile navel orange grown by grafting? Who would spurn receiving antibiotics (unnatural)? Who would want their children to be at risk to diphtheria, whooping cough, typhoid fever, or polio if those pathogens return and the choice is a dead or morbidly sick child or a child who has been vaccinated? Virtually all of our foods, traditionally bred as well as genetically modified, are unnatural because they have been shaped by human ingenuity and needs.

Supporters of genetically modified foods point out that rice modified to produce increased amounts of vitamin A will greatly reduce blindness (the vitamin A deficiency variety) in developing countries. Modified grains with a higher iron content will also diminish malnutrition in regions where the diet is poor in iron. Grains can be modified to produce more specific amino acids that are low in natural grains, and by those means, a diet based on grains can be as effective as one that uses meat. Malnutrition is a major killer of children, and it is ironic that where it is needed most (Africa) there is the greatest resistance because of a fear of genetically modified foods as toxic products being dumped by the U.S. because no one wants to use it in the U.S. (a claim that is false). Supporters also point out that genetically modified foods are in the earliest stages of development, and many new varieties will be constructed that will vastly increase farming in areas hostile to normal seeds (think of high salinity, flooding, arid climate, and fungal blights).

I believe that the progress of introducing genetically modified foods, which has been slow, is headed to eventual universal acceptance. Impediments are often political and economic. Countries with nonmodified maize will use the fear of new varieties produced by recombinant DNA technology and discredit the new strains as "Frankenfoods." The name implies that the food is toxic or harmful to health. Companies that manufacture genetically modified foods would prefer to have their products not labeled as such and mixed with varieties that are not modified.

Self-interest and corporate interest by farmers can lead to powerful opposition by consumer advocacy groups that see such behavior as evidence that something shoddy is being promoted and that its promoters desire a government sanctioned cover-up. Instead of honest debate, polemic arguments are substituted. Each side sees its motivations as directed for the public good. In expressing my optimistic view, I do not wish to whitewash corporate greed, pressure tactics, or lobbying efforts that favor companies manufacturing genetically modified foods. Those are very different issues from the safety, health benefits, and social good that genetically modified foods can provide, especially in parts of the world where malnutrition, if not starvation, is a real and chronic problem.

The major opposition today is in the European Union, bolstered by sympathy for organic farming by England's Prince Philip, who sees this form of agribusiness as the wrong way to solve the world's problems with malnutrition. Although there was an initial outpouring of resistance to genetically modified foods in the U.S., their usage has increased, and in the absence of harm, most people eating foods from their supermarkets are ingesting a significant amount of genetically modified foods.

We should remember that the 1970s and 1980s saw passionate debate over recombinant DNA technology, and some legislatures (even Boston's) sought to ban experimentation using these tools as too dangerous for its surrounding neighborhoods. In the 30 years of producing pharmaceuticals and hundreds of commercial and medical applications of recombinant DNA technology, there have been no runaway disasters because the slow, regulated, pace of development and use of that new technology (initiated by scientists themselves) have proved that the procedure is safe when properly regulated.

The important feature of the debate over genetically modified foods is the issue of regulation and testing.[10] That accompanying regulation is essential. It is almost impossible to test for imagined ills that do not exist, but it is possible to study products as allergens, and it is possible to prevent ecological damage from an unintended transfer of genes to the inappropriate species. It is also possible to study potential toxic effects on natural visitors to farms such as birds, butterflies, and bees. Such regulation is fair and reasonable. In all likelihood, genetically modified foods will be universally used, because their use by consumers shows no harm to health and because their planting and harvesting occur without harm to other crops. If "superweeds" do not emerge from these plantings, the fears will recede, as they did with recombinant DNA technology in the pharmaceutical industry. The use of completed genome sequencing with recombinant DNA technology will lead to innovations in food varieties, pesticide resistance, and growth on once incompatible soils. That makes a lot of sense to scientists. Overcoming public fears, however, requires a strategy of education and acceptance of regulation that is more effective than calling one's opponents Luddites.

Regulation may solve the issue of safety, but it does not address the ongoing problems of farmers around the world who have had to face competition when new technologies rendered their old ways of doing things unprofitable. Plows pulled by oxen gave way to mechanized tractors. The 40-acre farms were driven out, long before genetically modified foods were even theoretically possible, by the efficiency of 1,000- to 10,000-acre farms that are heavily mechanized. You can grow tomatoes in your small plot in the backyard of your house, but those will factor out economically for your investment of equipment and labor as $80 tomatoes. To buy tomatoes at less than a dollar a tomato, you need to buy them from supermarkets that deal with the more cheaply produced vegetables of mega-farms. Regulations for health and safety also do not deal with issues of corporate greed, and lobbyists for the opponents of genetically modified foods would have to seek legislative action to restrict practices that would force small farmers out of business. For almost a century, the U.S. has relied on subsidies to bail out unprofitable farms. Despite those efforts, family farming has diminished, and most farm families of mid-20th century America have disappeared as their children have sold their farms to agribusinesses.[11] In the U.S. farm communities, children have many options, but that might not be true in other countries, especially developing ones. In a world where each nation sees itself as serving its national interests, the chances of benevolent economic outcomes for the small marginal farmers around the world is slim.

Notes and References

1. Elof Carlson, *Mendel's Legacy: The Origin of Classical Genetics* (Cold Spring Harbor Laboratory Press, Cold Spring Harbor, New York, 2004). See Chapter 10, "The predominance of plant breeding to 1910."
2. G.H. Shull, "A pure line method in breeding corn." *Proc. Am. Breeders Assoc.* **5** (1909): 51–59.
3. E.M. East, "The distinction between development and heredity in inbreeding." *Am. Nat.* **43** (1909): 173–181.

4. A. Müntzing, "Genetics and plant breeding." In *Genetics in the 20th Century* (L.C. Dunn, editor, pp. 473–492). (Macmillan, New York, 1951).

5. M. Demerec, "Development of a high-yielding strain of Penicillium." *Carnegie Inst. Wash. Year Book* **44** (1945): 117–119.

6. A.F. Blakeslee and A.G. Avery, "Methods of inducing chromosome doubling in plants." *J. Hered.* **28** (1937): 393–411.

7. J.P. Mowrer, S. Stefanovic, G.J. Young, and J.D. Palmer, "Gene transfer from parasitic to host plants." *Nature* **432** (2004): 165–166.

8. J.D. Watson and J. Tooze, eds., *The DNA Story: A Documentary History of Genetic Cloning* (W. H. Freeman, San Francisco, 1981).

9. E.A. Carlson, "Genetically altered organisms: When the old becomes the new." *Dissent* (Winter 2001), pp. 56–61.

10. Peter Pringle, "Food, Inc." In *Mendel to Monsanto—The Promises and Perils of the Biotech Harvest* (Simon and Schuster, New York, 2003).

11. My wife's families were farmers or descendants of farmers in northern Indiana. Her ancestors began farming there before Indiana was admitted as a state in 1816. While visiting these farm areas in 2006, with my 92-year-old mother-in-law pointing out the old homesteads, I learned that each of them had been sold and that virtually none of the hundreds of descendants of her relatives, who were farming in 1914 when she was born, are in farming today. The small farms have virtually disappeared in Indiana.

PART 6

BASIC AND APPLIED RESEARCH IN MEDICINE

THE APPLICATIONS OF SCIENCE TO HEALTH vary in intent and quality. Medical research frequently involves human subjects, and that often follows after extensive trials with smaller mammals. Sometimes, however, the study is directly applied to humans, as in the Tuskegee study of tertiary syphilis. Some believe this experiment was similar to Nazi medicine described in Chapter 4, but I would not judge it so. I look upon it as flawed in execution, deficient in the standards we expect today (especially in the way whites looked upon blacks), and reflective of an older way of practicing medicine (paternalism) which we have rejected for the past 25 years.

I contrast that historical event with three ongoing fields of medicine and human reproductive biology. Cloning is both a red herring issue that brings out spurious science fiction scenarios of our future as a species and a serious effort for using stem cells (formerly called inner cell mass or preimplantation embryonic cells) as a treatment for numerous disorders involving cell loss or damage. In contrast to the syphilis study, issues of cloning and stem cells bring out religious opposition to both the research and its applications.

The related fields of in vitro fertilization and other forms of assisted reproduction also arouse religious opposition because their methods are considered by some as contrary to natural law theology and by others as inherently evil because frozen embryos are destroyed, intended for scientific research, or donated to other couples, all of which, for a variety of religious reasons, are considered by those critics to be an inappropriate response to those who experience infertility.

The last of these four medical examples involves prenatal diagnosis. In addition to opposition by religious critics who interpret this process as a "search and destroy mission," with abortion as its intended response to a disorder detected in an embryo or fetus, there are many who, for nonreligious reasons, distrust or oppose prenatal diagnosis because it looks to them like "eugenics through a back door."

Scientists prefer to ignore religion and believe it belongs to a separate domain of human experience and, thus, it is inappropriate for scientists to discuss religion. But when religious arguments are used, and when some 80% of humanity has strong religious beliefs, it would be folly for scientists to ignore this 80% of the electorate which often determines the composition of the government legislatures and its agencies, including those that scientists depend on for support. I believe the alternative to denial and avoidance of religious criticism is reasoned argument. Scientists can and should support their points of view and defend their fields when they are criticized.

12

Medical Deception and Syphilis

SYPHILIS IS A SEXUALLY TRANSMITTED DISEASE caused by a bacterial spiro-chete, *Treponema pallidum*. When it first appeared as an epidemic in Europe in 1494, it was a virulent disease that killed many of its victims within a year or two after infection. By the early 1500s, it had become attenuated (i.e., lost some of its virulence) in its passage through human bodies and was identified as a sexually transmitted chronic disease instead of a pestilential pox. Syphilis got its name in 1530, and it was treated until the 20th century by mercury compounds, which were highly toxic, especially to the nervous system. Most historians of medicine believe syphilis was picked up by Columbus's crew and brought back to Europe. Whether it had an Old World or New World origin, the outbreak that appeared after 1492 was new to Europe.[1]

Until 1838, many physicians believed that syphilis and gonorrhea were the same disease. The three stages of syphilis were worked out that year by Philippe Ricord. The bacterial cause of the disease was isolated in 1905, and an immunological test for syphilis was introduced the following year by August von Wassermann. A more effective treatment using an arsenical compound, arsphenamine (also called 606 or salvarsan), was developed by Paul Ehrlich in 1909. Both mercury and arsenic had drawbacks as medications, but the only improvement until World War II was a treatment using high fever, which was developed by Julius von Wagner-Juaregg, who induced malarial infections in syphilitics and then treated them with quinine after the *Treponema* spirochetes were supposedly killed by the fever. The fever treatment, like arsenicals and mercury compounds, was not an ideal treatment. One major drawback of the salvarsan treatment was its duration—intramuscular injections of the arsenical drug over one year. The treatments were painful, and for patients it was a personal and economic inconvenience to schedule reappointments for more shots. Things changed dramatically toward the end of World War II, when penicillin was shown to be effective for many bacterial disorders. Its effect on syphilis was remarkable. *T. pallidum* was highly sensitive to penicillin injections, sometimes disappearing from the tissues after a single injection.

When sexual intercourse takes place and one partner has an active case of syphilis, it takes about three weeks for the infection to manifest itself in the newly infected person.[2] If that person is a male, the penis (usually the glans) shows an ulcer-like sore. It is called a chancre and represents a collapsed pustule that has hardened. In a female, the chancre is usually found in the labia or the vaginal lining. The chancre disappears within a week or a little over a month. Some six weeks or so after its disappearance, this primary stage of syphilis is replaced by its secondary stage. A rash appears, often with localized erosive spotting (described as nickel-and-dime-sized lesions). The secondary stage lasts about two to six weeks, and then the person enters a latent phase when no symptoms are present. This latent stage may last many years, and some people (about one-third of all cases that are untreated) never develop other physical defects even into their old age. Most, however, enter a third stage of syphilis, and this tertiary stage is variable. Some show large ulcerations of the muscle, skin, or bones (called gummae). Some develop lesions in the vascular system and develop aneurysms. Some have a neurological defect leading to a staggering gait and paralysis (locomotor ataxia). Some develop a dementia as the brain becomes infected (paresis).

Syphilis can be transmitted by a mother to her unborn child, and such congenital syphilis may result in children with notched incisors and damage to the auditory nerves. Before the 20th century, such births were common, but in the penicillin era, they are very rare. There are some unresolved medical issues about syphilis. In general, most men and women in the later latent stage are noninfectious, but they usually test positive for antibodies in their immune system. When such patients enter the tertiary stage, it is not certain whether the damage leading to gummae and other organ disorders is caused by autoimmune responses or a reexpression of the once latent spirochetes.

The Rationale for the Tuskegee Syphilis Study

The U.S. government took an active interest in syphilis during World War I because 170,000 draftees, otherwise healthy, were in the primary or secondary stages of syphilis. The U.S. Army set up special treatment centers where the inductees received salvarsan treatments (an accelerated program taking several weeks of arsenical injections instead of one year). After the war, these treatment centers were disbanded, but interest in containing or eliminating syphilis in the American population remained, especially by the U.S. Public Health Service (USPHS).[3] It should be kept in mind that between 1910 and 1945 when syphilis was being treated with salvarsan, American society was divided on whether syphilitics should be treated. Many, for religious reasons, felt they should suffer the pain, stigma, and debilitation of their ailment because they had lived an immoral life. This was not related to race. It was a carryover of Puritan ethics, as it was called in the first half of the 20th century. I recall reading a case of a tenured professor in New York City about 1948 who was fired from his university position when his physician (as required by law) reported his name, affiliation, and address to the Public Health Service as having been treated for syphilis. Moral turpitude, in those days, was one of the few reasons a university could use to dismiss a tenured professor. Having syphilis (even if cured by prompt treatment) was prima facie evidence that the professor (in those days) was unfit to teach young people.

Eradicating syphilis was a difficult task, but the USPHS devised some programs in the late 1920s to begin that project. It was easier to identify and treat syphilis victims and their partners in urban settings where access to clinics was easier. In the rural South, however, this proved impracticable because of a lack of public transportation, a high incidence of illiteracy (especially among the predominantly black farmworkers in the cotton fields), and the poverty that existed before the Depression and even more profoundly during the Depression. The Tuskegee syphilis program began as an effort to eliminate syphilis. It was initially funded by the Julius Rosenwald Foundation of Chicago. They proposed working with the USPHS to improve the health of black working people in the South. The director of the program, Dr. Taliaferro Clark, suggested an eight-month study beginning about 1929, followed by the treatment of the day (salvarsan, mercury, and bismuth). However, the Rosenwald Foundation lost appreciable capital after the stock market crash of 1929 and had to withdraw from the study, leaving no funds for the treatment portion of the study.

To salvage the study, Dr. Clark suggested that the proposal be modified to use the county with the highest known incidence of untreated syphilis, Macon County in Alabama, for a special control study.[4] He estimated that 35% of the black population was infected. Dr. Clark envisioned following a select number of syphilitic and non-syphilitic black men in that county for the consequences of untreated tertiary syphilis. He selected tertiary syphilis because this was largely an unknown aspect of syphilis studies. In urban settings, persons showing gummae or paresis or other organ involvement would certainly see a physician, and they would be treated as best as could be by salvarsan or even by malarial infection, although no one knew whether those treatments at that stage were of any benefit. This control study would show whether the treatments helped. Syphilitics in the tertiary stage were associated in the public mind with the staggering movements of locomotor ataxia or the devastating senility of paresis. The dementia experienced by gangster Al Capone was etched in the public mind as the fate of those who were untreated until they entered the tertiary stage. (Capone did not die in prison. He was released in 1939, treated at a Baltimore hospital for paresis, and died in 1947 in Miami Beach.)

A second motivation for Dr. Clark's study was the possible difference in response to tertiary syphilis between blacks and whites. If syphilis had an Old World origin and it had been endemic in Africa, one could argue that blacks might have a milder course of the disease or that tertiary effects might be less severe. If syphilis had a New World origin, one could make the opposite argument that it might be more severe among blacks than whites, because whites were exposed to it for more than two centuries before blacks were brought to North America in the slave trade. There was also a lot of similar debate about malaria among blacks and whites in the South during the 19th century.[5] Blacks seemed to have a milder course or immunity to malaria, which took a heavier toll on whites living in the swampy parts of the South where malaria was prevalent until the 20th century. It was not known then that blacks had a conferred genetic immunity from a sickle cell mutation that made heterozygotes run a milder course of malaria. Sickle cell anemia was unknown to physicians before the 1950s. In addition, the milder response to malaria by many blacks made von Wagner-Juaregg's treatment questionable for use on black patients.[6]

This was the beginning of what became a notorious controversy in 1972 when the story broke about the government involvement in the Tuskegee syphilis study. The study involved five agencies—the USPHS, the Alabama State Department of Health, the Tuskegee Institute, the Tuskegee Medical Society, and the Macon County Health Department. The project began in 1932 and ended some 40 years later when news coverage led to a public outcry against the project. Dr. Clark retired and left the project a year after it began. In 1972, it was interpreted, largely by the press, as secret, racist, genocidal, an American version of Nazi medicine, and immoral.[7] In 1932, it was seen by its supporters as public, scientific, humane, ethical, and educational. Although a dozen scientific papers were published on the project in medical journals read by some 100,000 physicians throughout the U.S. during those 40 years, virtually no physician, white or black, had raised any objections to the study until the 1972 news coverage. How is it possible for the same study to be perceived so differently after 40 years?

Many issues have to be sorted out to understand this difference and to evaluate the motivations and issues in this controversy:

- American society was more overtly racist in the 19th century and the first half of the 20th century than it was in the second half of the 20th century.

- The revelations of the Holocaust and Nazi medical experiments during World War II created a public indignation about medical experimentation using human subjects without their consent. These revelations led to national as well as international laws governing the use of informed consent and the use of human subjects for medical research.

- The medical treatment for syphilis was less effective and more difficult to carry out before penicillin was introduced.

- It is difficult to know whether the informed consent given by the participating blacks had any meaning if the subjects giving it were illiterate, rural, and essentially cut off from general knowledge.

- Medical ethics in 1932 was steeped in paternalism; that attitude did not change until the 1970s.

- It was common practice for scientists working with human subjects to use deceit and to withhold information in the interest of having an unbiased control sample.

- It was common practice for human subjects who were institutionalized to be used in experiments without their consent.

- It was common practice to recruit prisoners hoping for reduced sentences without revealing the full risks of the experiments.

- The coalition of groups supporting the research included respected members of the black community.

There had been an earlier retrospective study, in Norway, of untreated whites who had tertiary syphilis. Dr. E. Bruusgaard in 1929 reported in a German journal on several hundred men in Oslo who had been diagnosed but not treated from 1891 to 1910 (neither salvarsan nor the malarial treatment was available then, but the more toxic mercury and bismuth treatments had been used for over 200 years).[8] There was no control for the study, but the Oslo males had a much higher incidence of cardiovascular complications than neurological complications. For Dr. Clark, this was an important possible difference that might emerge in a comparison of white and black untreated syphilitics. In Dr. Clark's mind, his study would be superior because it was a prospective study, following a cohort of tertiary syphilitics to their natural deaths while also following an equivalent cohort of non-syphilitics. In earlier studies, such as the Oslo survey, it was a retrospective study without an uninfected control, and the alleged excess of cardiovascular deaths might not have been caused by the tertiary syphilis.

We can ask ourselves at this point about the good intentions of the participating physicians and scientists. They saw themselves as objective, learning from this study the following information:

- The percentage of those in the later latent stage that eventually manifest tertiary syphilitic organ damage versus the percentage that die of natural causes showing no evidence of syphilitic lesions.

- The difference in site of organ damage among black syphilitics in the Tuskegee group compared to the white syphilitics in the Oslo study.

- The difference in intensity of expression of tertiary syphilis in black untreated men compared to white untreated men.

We could also ask a pertinent scientific question about this protocol. Why was there no third cohort selected, who would receive the salvarsan treatment? One reason was that it would have been hard to devise a way to keep the treated and untreated syphilitics ignorant of who had received treatment and who had been given some sort of placebo. The treatment choice in 1932 would have been to set up the intense venereal disease centers that had been used during World War I or to use the standard one-year course of treatment. Both would have been expensive to maintain. Isolating people for a month or more in an era with no unemployment benefits would have been both difficult and expensive. During the Depression years in the U.S., poor people suffered even more because they were unable to pay for their illnesses, take unpaid time off for prolonged treatment, or travel to get help from charity clinics. Blacks in the rural South would have been untreated with or without the existence of the Tuskegee project. If funds had been available and if the one-year method had been chosen, there was a fear of lack of compliance because the shots were painful and it was well known that compliance was difficult to maintain outside the enforced regimen of a military treatment center.

It is certainly true that white physicians in Alabama where the study was done were likely to share racist views about blacks.[9] It seems unlikely, however, that this program had any genocidal intent or reflected an utter indifference to blacks who were syphilitic. At this stage of the Tuskegee study (1932), one could fault its planners for a lack of zeal for treating people who in all likelihood would not have been treated in Macon County at a time (the Depression) when immense numbers of people, white and black, were cut off from medical attention because they had no money. They chose instead to use a bad economic situation for scientific gain. This might be a sloppy use of utilitarian ethics, but it was not malignant in the sense of Nazi doctors injecting typhus into concentration camp inmates to test new drugs. To get the subjects they needed for the Tuskegee experiment, the consortium came up with incentives. These included free examinations, free meals while being tested, free transportation to the hospital associated with Tuskegee Institute, and a modest sum (but a reasonable amount in 1932) of $50 for a contribution to their funeral.[10] The experiment at first involved 399 males with tertiary syphilis and 201 males who tested negative on a Wassermann test. A 1969 report summarizing the study listed 443 men with syphilis and 182 controls who were followed throughout those years. Of these, 36 controls (20%) were still alive and so were 53 syphilitics (12%). There were 40 infected wives (9%) and 19 children with congenital syphilis (4%).[11]

The subjects were not told they had syphilis (a term that was rarely used in public and probably not known to the men), but they were told they had "bad blood," a term that may have been used by those in rural areas who felt poorly. I suspect that this choice between educating the subjects on their condition and using a local term was based on the paternalism of the medical profession. The men were offered an unspecified "special treatment" in a letter entitled "Last Chance for Special Free Treatment" for their "bad blood." This treatment was actually a spinal tap that would search for evidence of neurosyphilis.[12]

The Tuskegee Experiment 1932–1945

Before penicillin, as we noted in our historical overview, the treatments were risky, but on a utilitarian ethical basis, both patients and physicians would rather go with arsenicals, mercurials, or malaria than with untreated syphilis for several reasons. First, treatment reduces the incidence of infections to others. Although it is true that most latent syphilitics are not infectious, that might not be true when tertiary symptoms appear if the spirochetes are suddenly reactivated in large numbers. The low transmission of syphilis from the tertiary syphilitics to their wives and children suggests that there may be natural immune responses against *Treponema* that develop on exposure in the vast majority of women and their children or that untreated syphilitics are usually noninfectious after the secondary stage. The interpretation of these frequencies was one of the unknown factors at the time of the Tuskegee experiments. Second, gummae, paresis, and locomotor ataxia, the symptoms associated with tertiary syphilis, are disfiguring or incapacitating. They can also lead to a premature death. Third, it is difficult to imagine a physician not treating patients when an effective treatment is available because it is the physician's duty to do so. Some of the participating physicians, however, may have been so convinced of the worth of the prospective study that they set aside their ethical duty to treat their patients when treatment was available. This suggests that some persuasive arguments were made to allow the tertiary syphilitics to go untreated. I suspect that the motivations of these physicians before penicillin became available were based on what they knew or believed at the time— the available treatments were good for primary and secondary stages but not for tertiary or late latent stages.

What made me lean toward that interpretation was a reading of *The Youngest Science*, a memoir by Lewis Thomas.[13] He described the day in 1937 when, as a young physician, he first witnessed a patient recovering from pneumonia at Bellevue Hospital. The patient had been given sulfa drugs, the first effective treatment for that disease. He described it as the birth of medicine as a science. Before that, he argued, physicians rarely treated diseases; patients cured themselves because the treatments were of dubious value. What treatments they received were primarily placebo in effect. If Thomas's assessment of medicine in his generation is correct, most physicians at the time of the study's inception then did not feel they were violating their oath for withholding a placebo or dubious treatment. I also believe they thought

Clark's design would guarantee controlled scientific data they did not have without actually doing any more harm than treating the patients in the tertiary stage when the medication was probably worthless.

The Introduction of Penicillin in 1944

As the war came to an end in 1945, a remarkable new drug was being used to treat bacterial infections. Although Alexander Fleming had found an antibiotic activity on petri plates contaminated with *Penicillium* mold and believed there was a product released by the fungus to destroy the bacteria on the plates, he was unable to isolate it or to produce it in clinically useful amounts.[14] That effort was achieved by Howard Florey and Ernest Chain, who grew *Penicillium* mold in flasks and decanted and purified the penicillin. The effects on infections were dramatic. Effective amounts of penicillin required some genetic engineering, and Milislav Demerec contributed to that in 1945 when he induced mutations using X rays to boost the amount of penicillin from selected strains.[15] With commercial production assured, penicillin rapidly became the agent of choice for many bacterial diseases—pneumococcal pneumonia, staphylococcus wounds leading to septicemia, gonorrhea, and syphilis.

Syphilis (primary and secondary stage) quickly became a treatable disease with minimal cost and inconvenience for the patient. There was some debate about the effectiveness of penicillin for tertiary syphilis. Some argued that by this stage the problems were those of autoimmune disease. There was evidence that in rats with neural syphilis (the equivalent of paresis), brain matter could be injected and cause syphilis in other rats; thus, the tertiary disease was not exclusively an autoimmune response and active spirochetes were present.[16] At the very least, the penicillin would arrest the progress of that tertiary disease. Any damage already done to the aorta, bones, muscles, nervous system, or other tissues was irreversible, but penicillin could be given to keep the condition from worsening.

This raises a moral question of greater significance than we can assign to the first (planning) or second (1932–1945) stages of the Tuskegee experiment. Why didn't physicians stop the experiment and treat the patients with penicillin? Is this not the key moral issue in the Tuskegee experiment? Our temptation is to assume that physicians quickly shifted in attitude to treating all stages of syphilis with penicillin. This may not have been true for skeptics who believed tertiary syphilis was an autoimmune disease and penicillin would have only a placebo effect, if any. A second factor may have been inertia or what is sometimes called "tunnel vision" in the popular press. Those committed to the outcome of the experiment may have felt that their patients, now elderly, would be of more value to society by running the course of the experiment and contributing to scientific knowledge than to a still uncertain claim of effective treatment for tertiary syphilis. It is not too different from the reasoning of a compulsive gambler who has lost too much to quit. That is a different weighting of utilitarian ethics than from the patient's perspective. Patients are more likely to gamble with an unproven treatment than with no treatment, but that does not make them morally superior for their choice. It is often the opposite feeling that is invoked when quack medicines are offered to the terminally ill who are desperate for any treatment; we then claim that the moral issue is exploitation of the dying by the greedy.

The Cincinnati Radiation Study Has Parallel Implications

One could fault the planners of the Tuskegee study for not aborting the project altogether in the absence of adequate funding to carry out their original plan. It is exploitive to use humans as guinea pigs to gain knowledge without a more forthright explanation to the recruits of what will be done to them. But that would have been asking people in the 1920s and 1930s to think like people today. Even after World War II, experiments were carried out in Cincinnati on terminally ill cancer patients who were recruited and given near-lethal doses of whole-body radiation so that military intelligence could gain information on the mortality risks from the damage associated with radiation sickness. The original aim of the experiments, carried out between 1960 and 1972 at the Cincinnati General Hospital of the University of Cincinnati, was to use whole-body radiation at high doses (but less than lethal doses) to see whether this method of treating terminal cancer patients was better than chemotherapy or no treatment at all. Interest in these experiments was expressed by the Department of Defense, which provided the funding for the research. This was the Cold War era, and the Defense Department was interested in the tissue damage and organ damage associated with high doses of radiation. Doses at Hiroshima or Nagasaki were estimates. The Cincinnati study would provide accurate dose–response data.[17]

The patients selected were mostly black (51 of 87), but that was the proportion of black to white patients in what was largely a municipal charity hospital. The patients were told they were receiving an experimental treatment for their otherwise hopeless condition, but they were not told of the side effects of radiation sickness and they were not told that the work was being supported by the Department of Defense. This was a classic case of conflict of interest.

The case eventually went to court, and a mixed decision in 1995 was provided. The U.S. District Court ruled that the patients could sue on limited grounds. The patients had been deceived and the university had failed to follow the national guidelines on informed consent and use of human subjects. The court rejected claims of damage from the treatments because the results of the experiments showed no difference in the outcome of radiation treatment when compared to the outcome from chemotherapy. In both cases, there was a slight prolongation of life over untreated terminal cancer. However, the court recognized the moral issue involved in its decision. "The allegations in this case indicate that the government of the United States, aided by the City of Cincinnati, treated at least eighty-seven of its citizens as though they were laboratory animals...."[18]

This Cincinnati case, like the Tuskegee case, suggests to me that when we go back to the time when these different experiments were carried out, we should try to understand why the scientists did what they did. This does not justify their actions, but it certainly gives us insights into their motivations and the means to prevent similar errors or deception in the future.[19]

Assessing the Tuskegee Experiment

James Jones's book *Bad Blood* does an admirable job of exploring all aspects of the Tuskegee experiment. His readers often are more selective in their reading of his analysis, and many see the experiment in the 1972 perception of racism, genocide, and Nazi medicine American style. I believe this is because most people do not have a historical perspective of issues. They project the present into the past and cannot imagine an era when medicine was more hand-holding and diagnostic than curative. This does not make the planners or participating health staff in the Tuskegee experiment free of all blame. One could argue that the experiment should not have been done even if the arsenical treatments were not useful for tertiary syphilis, because the program was based on a deception or uncorrected illusion that the syphilitic patients were being treated for bad blood. The circumstances and results of the Nuremberg trials were widely publicized after the Holocaust was revealed to the public. This did not stop either the Public Health Service or the participating physicians from stopping the experiment. It would continue for 20 more years. This is very troubling. Even if the physicians believed that the experiment had a momentum and value if completed, we cannot invoke the arguments used for the first two phases of the Tuskegee experiment. It is this moral blindness that contributed to the government apology that President Clinton made in 1997 to the Tuskegee experiment survivors and their families.[20] It is one thing to acknowledge racism as a factor in the decision not to treat the participants in the South in the 1930s to the 1950s, but it is difficult to see (from the hindsight of our present values) why that moral blindness existed to the same degree in the federal government after the Nuremberg trials or why the largely nonracist physicians in the North (including black and Jewish physicians) reading about the preliminary results in medical journals did not react until the 1970s. In all likelihood, the civil rights movement of the 1960s helped to shift American and medical opinions on the use of human subjects for medical research.[21]

What we do not know is how the staff would have reacted if it had been proven without a doubt in 1945 that tertiary syphilis was immediately arrested, and no further damage would occur, if penicillin were given. I suspect some would have voted to continue the experiment and some would have voted to stop it immediately and give all the tertiary survivors the penicillin shots they needed. In both cases, in good conscience, they would have justified their decision by the weighting of their utilitarian ethics.

Notes and References

1. Claude Quétel, *The History of Syphilis* (Johns Hopkins University Press, Baltimore, 1990).
2. Anonymous, *Syphilis: A Synopsis* U.S. Public Health Service Publication Number 1660 (U.S. Government Printing Office, Washington, D.C., 1968).
3. James H. Jones, *Bad Blood: The Tuskegee Syphilis Experiment—A Tragedy of Race and Medicine*, p. 61. (Macmillan, New York, 1981).
4. Ibid., p. 54.
5. For an overview of the historical and evolutionary aspects of this relation, see Krishna Dronamraju, ed., *Infectious Disease and Host-Parasite Evolution* (Cambridge University Press, United Kingdom, 2004).
6. Jones, op. cit., p.18, argues that in the 19th century the debate about the milder course of malaria among blacks was used to justify slavery because hired white field laborers would more likely die working in those fields. Note the utilitarian ethic in this argument. At the time of the Tuskegee experiment, the argument among physicians would have shifted to questioning the wisdom of using malaria

as a treatment for blacks if it did not cause prolonged high fevers.

7. Ibid., p 12.
8. Ibid., p. 92.
9. Ibid., Chapter 2: "A notoriously syphilis-soaked race."
10. Ibid., p. 153.
11. Department of Health, Education, and Welfare. Public Health Services and Mental Health Administration, Centers for Disease Control, Venereal Disease Branch. 1969. Item from Record Group 442: Records of the Centers for Disease Control and Prevention, 1921–2002. Series: Tuskegee Syphilis Study Administrative Records, 1929–1972. NAIL Control Number: NRCA-442-TSS001-TSSTUDY, ARC Identifier: 261643 at http://www.arcweb.archives.gov/.
12. The USPHS was required by law to give some partial treatment to the subjects. They were given a two-week treatment of salvarsan. In this sense, the Tuskegee experiment was scientifically tainted and could not be compared to the Oslo study. (James Jones, lecture given June 7, 2006 at Stony Brook University.)
13. Lewis Thomas, *The Youngest Science: Notes of a Medicine-Watcher* (Viking, New York, 1983).
14. R.D. Coghill and R.S. Koch, "Penicillin: A wartime accomplishment." *Chem. Eng. News* **23** (1945): 2310–2316.
15. M. Demerec, "Development of a high yielding strain of Penicillium." *Carnegie Inst. Wash. Year Book* **44** (1945): 117–119.
16. Anonymous, op. cit., p. 8.
17. Similar Department of Defense-sponsored studies were also carried out by M.D. Anderson Hospital for Cancer Research (263 patients, 1950–1956), Baylor University College of Medicine (112 patients, 1954–1960), Memorial Sloan-Kettering Institute for Cancer Research (20 patients, 1954–1961), and the U.S. Naval Hospital (17 patients, 1959–1960). Final Report of the Advisory Committee on Human Radiation Experiments. Chapter 8: "Total-Body Irradiation: Problems When Research and Treatment are Intertwined." 061-000-00-848-9. U.S. Government Printing Office. http://www.eh.doe.gov/ohre/roadmap/achre/chap8_1.html
18. Case no. C-1-94-126. U.S. District Court, Southern District of Ohio, Western Division 874 Fsupp 796:1995. US Dist LEXIS 401. January 11, 1995 decision, p. 23.
19. Martha Stephens, *The Treatment: The Story of Those Who Died in the Cincinnati Radiation Tests* (Duke University Press, Durham, 2002). See also Anonymous, by the ad hoc Review Committee, *"The Whole Body Radiation Study at the University of Cincinnati,"* 1972.
20. For the full text of this apology, consult "Remarks by the President in Apology for the Study done in Tuskegee," Office of the Press Secretary, The White House, May 16, 1997, http://clinton4.nara.gov/textonly/New/Remarks/Fri/19970516-898.html.
21. Paul A. Lombardo and Gregory M. Dorr, "Eugenics, medical education, and the Public Health Service: Another perspective on the Tuskegee syphilis experiment." *Bull. Hist. Med.* **80** (2006): 291–316. Lombardo and Dorr argue that during the planning stage, Dr. Clark and his USPHS coworkers were strongly influenced by the eugenics movement and that racial differences were specifically sought for that reason. How much such beliefs played a role is difficult to assess. Eugenics was widely embraced in that era, but equally compelling were the known differences in response to malaria between white farmers and their southern slaves.

13

Prenatal Diagnosis and an Alleged Eugenics through the Back Door

PRENATAL DIAGNOSIS WAS INTRODUCED IN THE 1950S. The first invasive probes of the fetus were used to replace blood for infants threatened with a blood-type-incompatibility condition called erythroblastosis fetalis. That condition was later treated by injecting expectant mothers with an antibody against the Rh blood antibody.[1] At least physicians knew they could penetrate the amniotic cavity of a pregnant woman with a needle and the baby would not be damaged. That led to the idea of withdrawing fluid from the amniotic sac, culturing the cells, and looking at the chromosomes of the embryo. The process was called amniocentesis. Chromosomes were chosen for study because Down syndrome (then called mongolism) was found in 1957 by French scientists to be caused by an extra chromosome 21 (the actual number was assigned later when a standard method of describing chromosomes was worked out).[2] Very soon after that discovery, two more disorders (both rare and lethal) were found for chromosomes 13 and 18 when they were in triplicate condition. These were called trisomies. Also in the early 1960s, some sex chromosome abnormalities were found among adults with conditions called Klinefelter syndrome (XXY males) and Turner syndrome (single X females). Less serious for the future sexual development of infants were XXX females and XYY males. All the other possible full trisomies, with very rare exceptions, aborted. Geneticists call the event leading to extra or missing chromosomes nondisjunction (this is an improper segregation during meiosis with unbalanced numbers of chromosomes in eggs or sperm).[3] The cells formed by nondisjunctional events are described as aneuploid.

A few years after the first use of diagnostic amniocentesis, Swedish scientists found a way to stain chromosomes so that they revealed a banded appearance.[4] Giant salivary gland chromosomes in fruit flies had been used for some 30 years to describe exquisitely small and complex chromosomal aberrations, and this new technique, using a fluorescent multi-ring compound (quinacrine), was very effective in describing some major genetic events in human cells such as inversions, translocations, duplications, and deletions. At about the same time, the first biochemical studies of amniotic fluid were being used to diagnose fetuses with Tay-Sachs syndrome, Hurler syndrome, or a number of other disorders whose defects were worked out by biochemists in the 1950s and 1960s.[5] This extended amniocentesis from cytogenetic studies to a more global genetic diagnostic procedure.

Throughout the last decades of the 20th century, better staining methods (using immunofluorescent techniques) and better sampling methods (using chorionic villi instead of cultured amniotic fluid) were introduced. Today amniocentesis is a standard medical procedure for

- all aneuploid conditions

- all chromosome aberrations

- biochemical defects in which molecules of a by-product in a blocked pathway accumulate in the amniotic fluid

- biochemical defects detectable by the absence of activity of an enzyme normally present in fetal cells

- identification of DNA that contains a mutant lesion for a given gene

Opposition to amniocentesis for anything other than treatment of disorders was almost immediate. The main argument was based on religious or philosophical grounds—it allows a female to know the condition of her embryo or fetus, and if the news is bad, she might elect to abort it. The person discussing the procedures, providing a reason for prenatal diagnosis and presenting the options once the outcome is known, is called a genetic counselor. The genetic counseling field had developed in the 1940s before there was a way to ascertain a fetus's genetic condition. Sheldon Reed coined the term and established some rules for genetic counseling.[6] He said the counselor should provide information that the client needs to know in order to make decisions about her reproduction. The counselor should not be directive (the client is the one who makes that decision). Both of these guidelines are still part of the genetic counseling tradition.

Elective abortion was not legal in the U.S. until some states began to allow the procedure for a variety of circumstances. Eventually, the Supreme Court ruled in the famous *Roe v Wade* case that a woman had a right to choose maintaining or ending her pregnancy with relatively few restrictions.[7] Of course, then and now, most abortions were social abortions—based on the female's feeling unprepared for motherhood because she was not married, too young, too old, too financially strapped, or psychologically unhappy with the idea of being pregnant (e.g., due to rape or incest). Once abortions were legal, medicine stepped in and used prenatal diagnosis with amniocentesis as a way to monitor females at risk for delivering infants with birth defects.

The reasons for aborting embryos or fetuses with potential birth defects are based on the following arguments:

- The child may suffer pain or a poor quality of life.

- The parents may be psychologically overwhelmed by the demands of an early death or chronic illness with no hope of a cure.

- The fetus's condition may lead to complications of pregnancy, even threatening the mother's life.

- The financial costs to maintain the infant may be staggering.

- The siblings who are normal might end up neglected because of the needs for caring for a hopelessly ill child.

- Parenting should not consist of a life of disappointment and low expectations.

The reasons physicians are willing to abort an embryo diagnosed with a birth defect include the following:

- It will prevent depression or even suicide of the mother based on her psychological profile.

- It will prevent pain and suffering of an infant that we cannot treat.

- The parents are thoughtful people who have assimilated the knowledge we gave, and we respect their autonomy to make this decision.

In the debates over the morality of prenatal diagnosis with elective abortion, those who oppose abortion will try to respond to these concerns in some of the following ways:

- Innocent life, regardless of the quality of that life, must be respected.
- Give the child up for adoption.
- Participate in a support group of parents who are raising such infants or children.
- The sacrifices you make may be ennobling and reveal a deeper meaning of love and life.

For those who abort, there is (to their critics) the implication of their being selfish and desiring "the perfect child." Parents who abort the embryo or fetus diagnosed with a birth defect scoff at that suggestion and argue that they would not abort a normal or even a mildly impaired child (e.g., one with a reparable cleft lip and palate). They would argue that they are being asked to be martyrs for someone else's religious beliefs, and they have no desire to do so.

Eugenic Arguments about Prenatal Diagnosis

The moral arguments will no doubt continue into the indefinite future because those who do not see elective abortion (especially for medical reasons) as prohibited killing will not feel they have done a criminal act. Those with utilitarian ethics will invoke a greater good (for the parents) in the outcomes for the parents with or without the raising of a child with a serious birth defect. But non-moral arguments have been raised by opponents of prenatal diagnosis with elective abortion that are based on eugenic considerations. The procedure is sometimes described by its critics as a "search and destroy" mission. The implied destruction is looked upon as one that changes the composition of the gene pool or that practices eugenics through prenatal diagnosis.[8] It is not clear whether this alleged eugenic ideal is in the minds of the genetic counselors, the physicians who do the abortions, or the parents who choose elective abortion. In my experience, I have never met a genetic counselor who rejected Sheldon Reed's advice to avoid, in good conscience, directing the clients into a decision. Similarly, every physician in human genetic services with whom I have discussed this issue has denied practicing eugenics. They tell me they distrust eugenics because it has had a bad record of picking on people who are more likely to be healthy and just down in their luck socially. On those occasions where I have sat in on genetic counseling sessions, the idea of eugenics never came up among the unhappy couples trying to assimilate bad news. My sample is certainly small, but I have not read any articles in genetic counseling journals which indicate that this is a major concern of prospective parents. I did meet some people older than I (which means they grew up in the 1910s and 1920s) who took the view that they would rather let a sick infant with an untreatable condition die than see its life prolonged. Some even argued that "society" was better off if such infants died, but I am not sure whether that reflection was based on the economic costs to society or the eugenic benefit of evolution doing its thing. I have only found eugenic sentiments among some of my genetics colleagues who are not in the health fields.

Those who see prenatal diagnosis as having eugenic consequences are only partially right and mostly wrong. Let us assume prenatal diagnosis with elective abortion was universal, with no opponents to it. What would the consequences be to the population's gene frequencies? I would argue the following[9]:

1. For autosomal recessive disorders (those on chromosomes 1–22) that are lethal, sterilizing, or incapacitating, and whose victims thus are not likely to live to reproduce, there would be no net difference if the affected embryo was born and died or if it was aborted. This group involves the largest category of single-gene mutations in humans (85–90%).

2. For dominant disorders of all sorts, there would be a dramatic reduction in the incidence, and the reappearance would be limited to the mutation rate for those genes (usually about 1 in 100,000 to 1 in 1,000,000 births). Note that they can never totally disappear because genes will always mutate, and thus, new cases (unless prevented by universal scanning) of retinoblastoma, achondroplastic dwarfism, Marfan syndrome, Apert syndrome, and the like will always have the potential of arising every generation.

3. For X-linked recessive disorders, like Duchenne muscular dystrophy, hemophilia, or Fabry syndrome, there would be a relatively rapid reduction in conditions by about 50% each generation. After six or seven generations, such conditions would be reduced to approximately their new incidence as new mutations, just as is true for dominant mutations.

4. At present, polygenic disorders are difficult to diagnose prenatally. Thus, most mental retardation, psychoses, and cardiac anomalies

would not be picked up by amniocentesis. Until some means of identifying their genotypes arises, the eugenic consequences of aborting them is limited to speculation.

The only way a substantial decline in autosomal recessive conditions could occur is if prospective parents elected to abort heterozygotes (like themselves). This is very unlikely, and even if it were chosen as a eugenic ideal, some future century from now, it would take many generations to reduce the incidence of these mutations to their spontaneous mutation frequency. One reason for this is that we are riddled with heterozygous mutations that, if rendered homozygous, would produce the very outcomes parents dread. Muller estimated that each of us harbors in our genotype an average of eight heterozygous mutations.

Other Concerns about Prenatal Diagnosis

There are certainly medical and other reasons why prenatal diagnosis needs to be discussed with a genetic counselor. There is a small but real possibility that the fetus may move and become impaled by the amniocentesis needle and consequently be born with a birth defect (especially if the needle enters the fetal cranium). There is a somewhat elevated possibility that the withdrawal of amniotic fluid might induce an abortion. These raised odds are not a major concern; they occur in less than 1% of fetuses subject to amniocentesis. As in many situations of life, there is no perfect process that addresses our concerns, and we deal with these conflicting bad outcomes with our utilitarian ethics.

Another factor is psychological. It takes about 12 weeks before amniotic fluid is abundant enough to be sampled, and an additional 4 to 6 weeks to culture the cells and obtain the cytological or genetic results. This means the fetus is in its second trimester; some people feel uncomfortable electing an abortion that late. Chorionic villus sampling reduces the time by at least one month, but it has the disadvantage that it samples cells from an extraembryonic membrane and not from the fetus itself. This raises the probability that there will be a false-positive reading because the aneuploidy may have arisen mitotically and was restricted to just that chorionic tissue. The opposite is also possible—a false negative, where the chorionic villus tissue is normal but the embryo has a mitotic aneuploidy that is missed. Both of these events are rare, but they are part of the complexity of life which offers minor exceptions to virtually all expectations.

There is an alternative to amniocentesis, but it is expensive and has its own set of concerns. For those who feel abortion of a fetus is too much of a burden morally but who would not mind choosing what preimplantation embryo to place in their oviduct or uterus, there is a method to sample one of the blastomeres of the four-cell stage, and if that embryo is homozygous for the defect, it is not implanted. This means only heterozygous or homozygous normal preimplantation embryos are used. The technique of blastomere sampling falls in the category of assisted reproduction with its financial cost and heavy medical management. The ethical reasoning in this case is a utilitarian one. Personhood is seen as a process of becoming; it is not absolute and not identified with a genotype at fertilization. A later abortion is more troubling to the psyche than an early one. The logic runs like this: Which causes a parent more grief, a neonatal death or a first-trimester miscarriage? Which causes more grief, a first-trimester miscarriage or a skipped menstrual period in which the zygote aborted so early that a pregnancy was unnoticed? Very clearly, people do not plan funerals for a skipped period, but they might for a child who dies a week after it is born. As we invest more of our expectations in a pregnancy the further along it goes, the psychological burden of loss increases.

Notes and References

1. Erthyroblastosis fetalis or Rh blood incompatibility arises when a woman has Rh-negative red blood cells and her partner has Rh-positive red blood cells. The fetus, if Rh-positive, may be attacked by antibodies from the mother's blood. This is usually prevented if the mother is administered Rh immune globulin (RhIg). Before this preventive method, and in some instances where the mother has antibodies against Rh-positive cells, an exchange transfusion is done. For a history of Rh, see David R. Zimmerman, *Rh: The Intimate History of a Disease and Its Conquest* (Macmillan, New York, 1973).
2. J. Lejeune, M. Gautier, and R. Turpin, "Étude des chromosomes somatique de neuf enfants mongoliens." *C.R. Acad. Sci.* **248** (1959):1721–1722.
3. There are many books on human cytogenetics. A classic reference source is J. DeGrouchy and C. Turleau, *Clinical Atlas of Human Chromosomes* (Wiley, New York, 1977).
4. T. Caspersson, S. Farber, G.E. Foley, J. Kudynowsky, E.J. Modest, W.E. Simonsson, U. Wagh, and L. Zech, "Chemical differentiation along metaphase chromosomes." *Exp. Cell Res.* **49** (1968): 219. Caspersson's laboratory used quinacrine hydrochloride, which intercalates into DNA and produces a banding that enables cytogeneticists to locate breakage sites in chromosomal rearrangements.

5. J.B. Stanbury, J.B. Wyngaarden, and D.S. Frederickson, *The metabolic basis of inherited disease* (McGraw Hill, New York, 1960. The most recent edition was issued in 1983.)

6. Sheldon Reed, *Counseling in Medical Genetics* (Saunders, Philadelphia, 1955).

7. The *Roe v Wade* decision of 1974 gave women an unconditional right to terminate first-trimester pregnancies. The court allowed states to set restrictions on second- and third-trimester abortions.

8. Ruth Hubbard and Elijah Wald, *Exploding the Gene Myth* (Beacon Press, Boston, 1999).

9. For a more detailed genetic analysis of these categories of human gene mutations and how they would fare in a universal prenatal diagnosis program, see Elof Axel Carlson, *The Unfit: A History of a Bad Idea*, Chapter 20, "The future of eugenics." (Cold Spring Harbor Laboratory Press, Cold Spring Harbor, New York, 2001).

10. H.J. Muller, "Our load of mutations." *Am. J. Hum. Genet.* **2** (1950): 111–176.

14

Cloning, Stem Cells, Hyperbole, and Cant

JUDGING BY ITS USAGE, I WOULD GUESS that the most widely read novel by journalists is Huxley's *Brave New World*.[1] I must have been in my early 30s when I first read it, snobbishly refusing for many years to read a book that almost everyone else I knew had read. Part of my prejudice was my belief that the book was anti-science, and because I wallowed in the pleasures of learning science and seeing what good it has done in the world, I felt this was not worthy of my time. I learned later to appreciate the book and even assigned it in freshman seminars and as collateral reading in my non-majors biology courses. The book is not anti-science. It criticizes the abuse of science as a tool to establish and perpetuate totalitarianism. The book is more a denunciation of Mussolini, Stalin, and Hitler than it is a denunciation of science. I later learned that Huxley was inspired by J. B. S. Haldane's earlier nonfiction best-seller, *Daedalus*, one of the few books predicting the future that was on target most of the time.[2]

Huxley popularized the science-fiction genre of cloning. In this popular form, there is the persistent error that a person (including that person's personality, memories, and sense of self) can be cloned. That is certainly impossible by any of today's technology and may be one of the few things that is technically impossible (for starters, how would one program separately 100 billion neurons, each with its corresponding molecular inventory and unique set of dendritic filaments?). One would think that our familiarity with identical twins would establish the uniqueness of individuals, even when they have an identical genotype. I have known lots of twins because human genetics was one of the subjects I taught, and I have come to appreciate virtually all identical twins in my classes as sets of two different people. Sometimes the differences are striking. I met an identical twin at a genetics meeting. She was a professor in the Midwest and said she grew up in rural Georgia. While in high school, she liked to hang out with her fellow nerds and her sister liked to hang out with her fellow "rednecks." She said she and her twin sister have very little in common, and when they get together at holidays they differ in accents, interests, and values. I also know identical twins who are much alike, sound alike, and whom I have easily confused, but none have claimed to share each other's thoughts at will except for one pair I caught cheating on an examination. The one I had questioned said she was communing with her sister by thought transfer. I told her I would be only too happy to apologize if she and her sister would take a makeup examination sitting far apart in an empty room with me so they could commune silently across the room. They declined to add perjury to their academic sins.

In the popular image of cloning (as depicted by Mickey Mouse's mop in the sorcerer's apprentice sequence of *Fantasia*) one individual becomes multiple individuals. The self is cloned. I have never heard people wishing they had an identical twin of uncertain personality, career goals, and values. That would be the risk for those who choose what I call vanity cloning in contrast to the reality of natural twinning. My suspicion is that if artificial cloning is ever introduced, within a generation it will be a little-used fad because, more often than not, the hopes of the older twins will not usually be realized when they raise their younger twins. I doubt if would-be cloners have thought about the psychological problems they would experience. It is one thing to be a natural identical twin. You both don't know what you will look like some 10 or 20 years into the future. But imagine the trauma to a young male clone seeing the undesired features awaiting him as he realizes he will be bald, pot-bellied, and somewhat ugly in late middle age. And that's just the physical future. If personality is truly set by one's genes, it would be equally traumatic if one's twin parent is an unbearable bigot, domineering, and hypocritical.

Some Biological Aspects of Twinning

Contrary to popular belief, most twins are not formed by two cells falling apart after the first cleavage division of the fertilized egg. At most, 20% of identical twins arise at that time or prior to forming a blastocyst. The blastocyst consists of an inner cell mass (now called its stem cells) and a surrounding tissue that contributes to membrane formation, to attachment to the uterus, and to feeding the embryo but not to the formation of the embryo itself. The other 80% of identical twins form by a partitioning of the inner cell mass (they are now called the stem cells) of the blastocyst or at a time when that inner cell mass has differentiated into embryonic tissues (i.e., ectoderm, mesoderm, and endoderm).[3] This leads to a variety of patterns of extraembryonic membranes. If the two first blastomeres each form a separate blastocyst, each twin will have its own amnion and chorion, the two major membranes associated with mammalian pregnancy. If the morula (ball of cells) prior to blastocyst formation splits into two chunks, each will also form a separate blastocyst and, hence, each twin will have a separate set of amnion and chorion around it. If the blastocyst forms and the inner cell mass splits into two chunks, the most common event is two embryos, each with its own amnion but with only one chorion around them. This is the most common type of twin membrane associated with identical twinning. In general, those that partition later will have one set of membranes (amnion and chorion). This is relatively rare, and the two embryos or fetuses can come into contact with one another and twist about their umbilical cords, resulting in the death of both. Those with one chorion have a problem in addition to crowding, because their blood vessels commingle in the placenta, and this can lead to unequally nourished identical twins that look dissimilar at birth (or the smaller twin may have birth defects, a lower IQ, or a delay in catching up to the other twin). This is inequality of identical twins even before they are born. The least fortunate timing is a very late partitioning of the embryonic mass, because it can lead to conjoined twins.

Conjoined twins are the despair of theologians. A two-headed twin with an otherwise normal body presents problems on counting the child as one or two persons. A one-headed twin with two lower bodies is more bizarre but also presents counting problems. Even when twins are separate and look very much alike, how does one characterize the concept of personhood as beginning at zygote formation (a favorite theme of right-to-life advocates) if the two babies have a common zygotic origin? Do they share one soul or two in theological analysis? If they have separate souls, were they placed in them after the twinning event occurred, even as late as an implantation embryo with partially differentiated embryonic tissues? What was the personhood status of the blastomeres, morula, or inner cell mass before the twinning event occurred? Some scientists raise these questions to point out the complexity of biological life. They argue that religion-based deontological values often ignore this more complex biological reality.

Stem Cells and Medicine

The term "stem cells," about 50 years ago, was used in the histology text-books to describe partially differentiated cells. There were stem cells in the bone marrow that produced a variety of blood cells, red and white. In some animals, like salamanders, there were stem cells that formed after amputations of a limb resulting in a heap of undifferentiated cells (called a blastema then) that formed a new limb. Quite separate in those days were the cells of the early embryo after implantation, called the inner cell mass, from which the ectoderm, endoderm, and mesoderm formed and from which in turn the entire body of an individual developed. Today, the term stem cells has largely replaced these older and more specific terms. It also has created a lot of debate and confusion about their usage.

Before these cells were called stem cells, they were identified as having what embryologists called totipotency. Cells are totipotent if they can form all the other tissues of a body. In the 1890s, German scientists were inspired by Hans Driesch, who separated the blastomeres of a two-celled sea urchin and produced identical twins that were perfectly healthy and fertile. By 1914, Hans Spemann isolated a nucleus from a cell that was in a 16-cell-stage amphibian and inserted it into a set-aside mass of cytoplasm of that original fertilized egg. It produced an identical twin of the mass containing the other 15 cells. From this experiment, Spemann concluded that the nuclei of cells were totipotent at least up to the 16-cell stage.

The work of experimental embryologists until the 1950s did not advance in this field because it was technically difficult to isolate nuclei of cells. Robert Briggs and Thomas King changed that by using frog blastomeres and early embryo cell nuclei to produce embryonic clones. These differed from twins because they were not partitions of an embryo. They used a donor nucleus placed into a fertilized egg whose nucleus had been destroyed or removed. The donated nucleus could then be tested for totipotency. Briggs and King found that totipotency was the rule with normal embryos formed, until the ball of cells (called a blastula in amphibian embryos) shifts into a gastrula (the stage when ectoderm, endoderm, and mesoderm arise from the stem cells or totipotent blastomeres). The resulting embryos aborted or produced abnormal gastrula stages.[4] John Gurdon extended this work using toads and found that some of the gastrula tissue (especially the gut-forming endoderm) was still totipotent, but later stages were not.[5] By the 1960s, it was widely believed that gene silencing mechanisms during differentiation were the reason adult nuclei could not be used to generate clones. The work of Ian Wilmut's team in Scotland came as a surprise to most geneticists and embryologists. Dolly the sheep was a product of an adult nucleus from a sheep breast and the cytoplasm of an enucleated egg. The technique is difficult to do because only one of several hundred nuclear transfers works. Resetting the genes of the donor nucleus is a lucky event, and exactly when and how that happens is still not known.

How Stem Cells Became a Medical Interest

The hope of this new field of experimental medicine is to find stem cells that will differentiate in damaged heart muscle to restore a normally beating heart; to find stem cells that will commingle with cells in the brain to restore function to those with Parkinsonism, Huntington syndrome, Alzheimer syndrome, or even strokes; to repair ripped-up nerve bundles traumatized by broken necks and backs; to eliminate medication for some types of diabetes by providing stem cells that will differentiate into Islets of Langerhans capable of secreting a functional insulin. The list will no doubt grow as successes occur in what remains a field based on hope, logic, and some experimental support from both animal and human trials.[6] It is not a surprise that stem cells should be perceived as a medical panacea. If tissues are compatible (e.g., blood types, HLA types), they can be used to save lives. Even if they are not immunologically compatible, the field of transplant medicine has shown how parts of the immune system can be suppressed to allow grafted tissues to function without leading to death from infectious diseases.

By changing the name from inner cell mass cells (an awkward phrase) to stem cells, physicians shifted the interest from embryology to medicine. Stem cells are a source of new adult cells, and bone marrow transplants are stem cells that the leukemic individual uses to replace the diseased marrow that has been killed by high doses of radiation or by chemotherapy. The history of transplant surgery since the first heart and kidney transplants in mid-20th century taught us that progress is steady, punctuated with setbacks, and troubled by ethical conflicts and moral debates; but eventually it becomes a standard medical practice no longer of interest to its initial critics. Few people today invoke the 1960s playing-God arguments, the conspiracy theory (poor people will be murdered for their organs to sell to the wealthy), and the "who will be the lucky recipient" question (e.g., scenarios of the young prostitute drug user or the venerable aged bishop dying of kidney failure). Today, tens of thousands of people receive kidneys, hearts, livers, and other organs from those who die of accidents and untimely deaths.

Stem cells from bone marrow are not as plastic in their range of potential cell types as stem cells from a blastocyst. This raises some important problems. Blastocysts are readily available by the tens of thousands from clinics where in vitro fertilization (IVF) is performed. There are far more preimplantation embryos in liquid nitrogen than will ever be used by the parents whose gametes formed them or by infertile couples requesting their use to achieve a pregnancy. The donors of these preimplantation embryos have already signed consent forms for those cells to be used for research. At present, at least in the U.S., these embryos cannot be donated for research because there are federal laws preventing the funding of research based on such stem cells. Very clearly, utilitarian ethics takes a backseat to rights-based ethics on this issue. It's OK to justify dropping bombs on Hiroshima and Nagasaki to save lives (American and Japanese), but it is not OK to use stem cells from IVF repositories to save an even greater number of lives than the potential to form babies from these repositories. Rights-based ethics presumably apply to the unborn (i.e., innocent life) and not to the born, who are, regrettably, expendable in times of war (but so are the innocent unborn in those pregnant Japanese women crisped by atomic explosions, unless their being innocent is stripped by virtue of their being Japanese). Consistency is difficult to maintain in either rights-based or utilitarian arguments.

Both sides of this debate have their own problems with hyperbole. To convince opponents or the uncommitted that stem cells should be used for research and, if successful, on a large scale for medicine, are those who see almost every tissue loss and damage as repairable by an appropriate injection of stem cells. I am reminded of the hyperbole expressed by those who first advocated nuclear reactors to replace conventional fossil fuel power plants for generating electricity—it would be so cheap to produce that it would be virtually free. Equally optimistic are the veiled utilitarian arguments of political opponents of blastocyst stem cells who believe that the present supply in the U.S. (somewhere around 20 strains of useful ones) should be sufficient to carry out all the research one needs (note that the absolutism of rights-based ethics took a beating on that Solomon-like split-baby suggestion by the 2003 White House).[7]

The need for research in stem cells is essential. If the human stem cells from blastocysts are totipotent, there will be little if any gene silencing involved in those nuclei. That is very likely because of the way identical twins form during the first week or two after fertilization. But whether stem cells from a blastocyst will respond in the same way as young neuroblasts or other immature tissue in spinal cords, hearts, livers, and pancreases, is not known. Will they be restricted to just the tissue they are intended to replace, or will they, like teratoma cells, form undifferentiated masses of tissue (teeth, hair kidney cells, kidney tubules) inside the brain of a person with Parkinson disease? The only way to study that is through animal experiments and, eventually, human experiments. The process of any new medical innovation is slow, requires caution, and is subject to surprising findings not anticipated from animal work or computer models.

There are many possible ways this stem cell research will evolve. It may be possible to work out by biochemical means fully dedifferentiated stem cells from marrow or other somatic sources. Such cells may function like blastocyst stem cells (i.e., they would be totipotent, meaning they can form an entire embryo). Those who oppose stem cells from blastocysts have no problems with this if the stem cells are derived from somatic tissue. But what if those same somatic-derived stem cells are used to establish a clonal pregnancy? Does this mean that fully totipotent stem cells derived from somatic tissue, like stem cells from blastocysts, have personhood? Since they are derived from somatic tissue, does this means all somatic tissue is potentially infused with personhood? This is not as far-fetched as it seems. I am reminded of an experiment many years ago by Thomas J. King, who used nuclei from cells of a frog kidney tumor to produce a perfectly healthy cloned frog.[8]

The Motivations for Vanity Cloning

The desire for vanity cloning has both a serious and a frivolous motivation. The serious motivation is tied to positive eugenics and the hopes from Francis Galton's first proposals to H. J. Muller's final pleas for a eugenics based on individual germinal choice and steeped in values and traits he cherished—intellect, talents, health, and that type of leadership that brings out the best in others (he called this "fellow feeling").[9] Muller assumed, and had no evidence, that these socially admired traits had strong genetic underpinnings. Muller died in 1967. There is still no evidence in the first decade of the 21st century that these traits are heavily fated by heredity. Basing a eugenic program on hope alone is not a good idea. I was a student of Muller's and think the world of him as both a scientist and a person who cared about others, but over the years I have become less convinced that these traits (especially the behavioral ones) are primarily genetic. Within a generation, it is likely that many of the genes associated with our nervous system will be worked out, and we shall know then whether such traits do stem from a variety of fortunate and uncommon genotypes.

The appeal of vanity cloning for frivolous reasons is different from the idealism of positive eugenics. Here the motivation is self-image. Narcissism would be the term Freud would use. People do like themselves, and there will be some who look upon vanity cloning as a biological tool for immortality. If they are not deluded into believing that their personalities and potential for accomplishments will be cloned, they do believe that their genotype merits perpetuation not in any half-way measure through one's gametes, but in full measure through a somatic cell used to form a clone. If the day comes when such vanity clones are possible, there will be added heavy-duty psychological burdens on such children trying to live up to the donor's reputation. Vanity is not limited to males, so cloning can be for either sex. It is highly unlikely that the children of such experiments will also be enamored in the same way. An indefinite line of cloned twin descendants for unbroken generations for all or most of the vanity clones is far-fetched. Thus, vanity cloning would probably be mainly to please the donors of only one generation. That's not a long line of descent leading to the immortality motivating the cloners. If narcissism is more likely the reason for cloning oneself, this is not the most inspiring lesson for the cloned child to absorb.

Although the normal and behavioral concerns are sufficient to make a potential vanity cloning a decision that should not be hasty, there are biological features of the cloning project that make a present-day attempt risky. In most animals, as was true for Dolly the sheep, cloning is difficult. It takes hundreds of tries to get a clone to be born alive and survive to maturity. A major reason for this may be associated with the way genes are imprinted (chemically modified) by methylation during differentiation. Getting a truly totipotent cell from an adult somatic tissue is not easy. Setting the clock back to a zygote-like state may enable such a cloned baby to be born, but whether such an infant will have psychological problems (psychosis, mental retardation, autism, etc.) from the imperfectly erased imprinting is difficult to predict. Who would ethically take that risk, assuming the prospective parents are informed about it, if the experimental work is lacking to demonstrate that the process of forming appropriate stem cells from adult somatic cells is identical to that of forming zygotes?

The science fiction arguments for cloning have faded in the past 25 years. No doubt *Brave New World* inspired science fiction writers to portray armies of clones fighting each other. Why a cloned army would be superior to the work of a good drill sergeant to lockstep diverse troops into fighting fitness is hard to figure out. Wouldn't a variety of minds be better to fight a war than one group-mind thinking the same way and not knowing how to take advantage of unexpected surprises and circumstances? What happened to the virtues of variation in natural selection in the thinking of these cloning enthusiasts? Equally extreme and silly is the belief that cloning will lead to a massive loss of variation in humanity. How could that be if those who could afford it and would be narcissistic enough to see it through amount to no more than a few hundred people per generation? Cloning about 0.00000001 percent of humanity is not likely to cause a serious change in the world's gene pool.

Notes and References

1. Aldous Huxley, *Brave New World* (Doubleday Doran, Garden City, New York, 1932).
2. Krishna Dronamraju, ed., *Haldane's Daedalus Revisited* (Oxford University Press, New York, 1995). This book is a reprinting of Haldane's 1924 essay, with eight additional essays commenting on its significance.
3. M.G. Bulmer, *The Biology of Twinning in Man* (Clarendon, Oxford University Press, New York, 1970).
4. R. Briggs and T.J. King, "Transplantation of living nuclei from blastula cells into enucleated frogs eggs." *Proc. Natl. Acad. Sci.* **38** (1952): 455–463.
5. J.B. Gurdon, "The developmental capacity of nuclei taken from the intestinal epithelium cells of feeding tadpoles." *J. Embryol. Exp. Morphol.* **10** (1962): 622–641.
6. L. Walters, "Human embryonic stem cell research: An intercultural approach." *Kennedy Inst. Ethics J.* **14** (2004): 3–38.
7. Anonymous. 2002. Guidance for investigators and institutional review boards regarding research involving human embryonic stem cells, germ cells, and stem cell derived test articles. NOT-OD-02-044 National Institutes of Health, April 10, 2002, Washington, D.C.
8. T.J. King and M.A. DiBerardino, "Transplantation of nuclei from the frog renal adenocarcinoma. I. Development of tumor nuclear-transplant embryos." *Ann. N.Y. Acad. Sci.* **126** (1965): 115–126.
9. H.J. Muller, "Germinal choice: A new dimension in genetic therapy." *Excerpta Medica*, Abstract No. 294 (1961), p. E135.

15

Assisted Reproduction and the Argument of Playing God

AMONG COUPLES INTENDING TO HAVE CHILDREN, about 20% will not have conceived after one year and about 8% will not have conceived after two years. People who have trouble conceiving are called infertile. If they never have children without medical assistance, they are called sterile. Infertility can involve a male factor, such as a low sperm count, defective sperm morphology, or some biochemical defect in the sperm. About 40% of infertile couples have a male factor defect. Similarly, about 40% of infertile couples have a female factor such as nonovulation, blocked fallopian tubes, or some abnormality of the endometrial lining of the uterus. About 20% of infertile couples have both a male and a female factor involved in their difficulties conceiving. Until the late 1970s, most sterile couples were out of luck. They could remain childless, they could adopt, they could elect donor insemination (then still called artificial insemination), or they could elect female hormone injections to stimulate ovulation (such as the synthetic estrogen, clomiophene).[1]

Things changed in the 1980s when a variety of new procedures became available for infertile couples. A new field of medicine, assisted reproduction, emerged. The major techniques used for in vitro fertilization (often abbreviated as IVF) now involve:

1. Introducing sperm by catheter into an oviduct at the time the female is ovulating.

2. Removing eggs, fertilizing them with the male partner's sperm in a plastic dish, and introducing the preimplantation embryo when it is about 2–4 cells along into the oviduct (the classic IVF).

3. Introducing to the uterus a more advanced preimplantation embryo, about the time it forms a blastocyst (the unit that has an outer membrane and an inner mass of stem cells and which normally implants in the uterus).

4. Using an egg, usually donated from another female to a woman who cannot produce her own eggs, and fertilizing that egg in a dish with the sperm of the infertile female's male partner.

5. Using another female (usually a volunteer relative) to receive the fertilized egg of a couple where the female has no uterus or a defective uterus that cannot support a pregnancy.

6. Injecting a deformed or immature sperm into the egg to produce a fertilized egg (a process called intracytoplasmic sperm injection or ICSI).

When assisted reproduction first started, the techniques were still evolving and the rate of success was low (about 10% for a live baby).[2] Today, the techniques have much improved, and the chances of success hover about 30–40%, depending on the skills of the laboratory, the age of the women who seek pregnancy, and the choice of techniques used by the laboratory. There are many factors that determine the outcome:

• Women over 40 have a more difficult time than women in their 30s. Older eggs do not do as well. Some laboratories try to compensate by inserting cytoplasm from a younger woman's egg to rejuvenate the egg. The nuclear chromosomes are those of the infertile female, not of the younger egg cytoplasm donor. Older sperm is also less likely to result in a successful fertilization. Although some males can be fertile into their 80s, the sperm of most males over 50 becomes progressively more infertile with age.

• Women over 40 have a higher chance of producing eggs with extra or missing chromosomes, and most of those abort the embryo (very early in pregnancy).

• Little is known about the factors determining implantation and how to compensate for mutations that prevent successful implantation.

• Little is known about the genes involved in the first few days of pregnancy when the fertilized egg undergoes cleavage, forms a ball of cells, and then a blastocyst, or when that blastocyst forms embryonic tissues. The genes for these events are still to be located and functionally described from an analysis of the human genome project.

Biological Factors and Assisted Reproduction

Because the chances are low that any single fertilized egg introduced into a uterus or oviduct will lead to a live birth, those who practice assisted reproduction place more than one gamete or preimplantation embryo into the recipient female. This leads to a higher incidence of multiple births (nonidentical twinning). Multiple births have a higher incidence of birth defects and create more difficult pregnancies and deliveries. When four preimplantation embryos were the norm (usually 4-cell stage), about 75% singletons, 20% twins, 5% triplets, and less than 1% quadruplets would be born. Using blastocysts greatly reduces multiple births, because one or two are usually used for implantation. For reasons unknown or still speculative, however, there is an elevated risk of identical twins occurring.

Women who elect to have assisted reproduction usually have their pregnancies monitored because most are older females and they are at risk for abnormal chromosome numbers in their eggs and their potential infant. Prenatal diagnosis detects such at-risk pregnancies, and women usually choose to abort those embryos.

Women who go through the procedure of assisted reproduction have numerous injections of hormones to stimulate their ovulation at the appropriate time for an in vitro fertilization procedure. The long-term effects are still not known because the IVF procedures were not widely used until the 1980s, and it will not be clear until about 2020 whether these hormonal procedures have had any effect on elevated cancer risks or heart attacks (the most obvious side effects of introduced steroid drugs). The good news is that the children who have been produced by IVF are as normal as those who are born to fertile couples. There will no doubt be some males produced by ICSI (intracytoplasmic sperm injection) who will inherit their father's infertile Y chromosome defect.[3] Because most gene mutations for female factor infertility are likely to be recessive, the heterozygous daughters will probably be fertile, but may pass the infertility genes they got from their mothers to some future descendant that will express that infertility. This raises a question rarely discussed by critics of assisted reproduction. Is the process inherently dysgenic? Natural selection eliminates gene mutations that disrupt normal function, and sterility is one such dead end for mutant genes. Assisted reproduction should, in centuries to come, gradually elevate the percentage of infertility in the population. One reason this issue might not be raised is that all medical procedures that treat diseases and disorders are inherently dysgenic and we are (for good reasons) neither ready nor willing to consider eugenic measures to compensate for such accumulations of mutations.

Religious Assessment of Reproduction

Sex is a preoccupation of most religions, and when one thinks of sin and morals, one's first impulse is to think of sexual behavior. There are restrictions on marriage (most religions allow one male to marry one female), restrictions on when to have intercourse, restrictions on how that intercourse and intimate behavior should be done, familial restrictions on who can marry (decrees on consanguinity that are allowed or forbidden for marriage), and religious restrictions on who can marry (Orthodox Jews, for example, mandate their children not to marry non-Jews). The 20th century saw the introduction of two medical approaches to reproduction. The first was the birth control movement initiated by Margaret Sanger in the U.S. and by Marie Stopes in the U.K. That movement shifted from a criminal advocacy to an accepted norm in most of the world. All industrial nations use family planning with artificial contraceptive means to reduce the mean family size to replacement levels. Many countries, like China, have a one-child-per-family ideal. Russia has less than that. The number of children born to a family from 1900 to 1999 dropped among industrial nations from an average of four to an average of two.

Opposing that birth control movement was the Roman Catholic Church. Their objection is based on natural law doctrine. Under natural law, the purpose of sexual intercourse is procreation. Anything that thwarts that intent is in violation of natural law. There are few religions that base their religious views of reproduction on natural law. Protestants don't. Jews don't. But objections to birth control also include other moral arguments if the agents used prevent fertilized eggs from implanting or destroy such fertilized eggs or preimplantation embryos. This is where a coalition of Roman Catholics and fundamentalist Christians and Orthodox Jews can join forces in a Right-to-Life movement that sees such uses of contraception as acts of murder. Those who reject these views claim that murder applies to live-born individuals, and a fertilized egg does not have the same status as a live-born baby. A sick embryo does not have the same moral status as a sick baby needing treatment. We allow nature to abort the former as a spontaneous miscarriage but not to take the live-born without a struggle on our part to keep it alive.

The second medical innovation of the 20th century was the attempt to treat infertility by scientific medicine. This included hormonal methods in the 1960s, with success for those females who could not ovulate their immature eggs. Low sperm counts were compensated by concentrating the husbands' sperm samples and using what was then called AIH (artificial insemination by husband). Artificial insemination by donor did not solve the infertility problem for the infertile male, but it did give what was called a half-adopted child for the couple to raise. It was in vitro fertilization that changed fertility from a low to a high success rate with a new field of reproductive medicine assisting the infertile. Like the birth control movement, it has become universal in industrial nations and among the middle classes and wealthy wherever they are. Physicians see an obligation to the infertile as they do to children with birth defects. If it is treatable and the patients ask for help, they help. Virtually gone among the educated is the fatalistic outlook of the past ("this is your cross to bear").

The Moral Objections to Assisted Reproduction

When the first baby conceived by in vitro fertilization was born in 1977, in England, the world news response was sensational, and baby Louise Brown was called a test tube baby.[4] The team that brought her into being, Patrick Steptoe and Robert Edwards, was trained at the Jones Institute in Virginia, but they decided it would be too risky to attempt an IVF pregnancy in the U.S., especially if there were complications. It would have killed the field.[5] In one respect, the public excitement and commentary was due to novelty, as in the first heart transplant carried out by Christian Barnard in South Africa. There was also the added sexual component that never ceases to attract journalistic comment, because in vitro fertilization was first discussed by J.B.S. Haldane in the 1920s and became a centerpiece for Aldous Huxley's *Brave New World*, a novel that continues to have a powerful impact on advances in biomedical sciences. Huxley's objections were very different from the religious objections to in vitro fertilization. He foresaw a dystopic world in which science is used as a tool to regulate human behavior and to structure society with a totalitarian efficiency. It was an era still attracted to the eugenics movement and the belief that unfit people should be sifted out by a state control over reproduction (marriage laws, compulsory sterilization laws, and a growing sympathy in Germany for race hygiene).

Those social objections were overshadowed, as I have pointed out, by religious objections, especially from the Roman Catholic Church. In Catholic doctrine, reproduction had long been regulated. Birth control was limited to the rhythm method, in which no chemical or mechanical procedure is introduced to prevent gametes from uniting and no interruption of the sexual process takes place. This made any chemical prevention of pregnancy by hormones or by destruction of gametes nonpermissible. At issue was not the killing of gametes but the violation of natural law doctrine. Before there was a cell theory (1838), such ideas did not exist, and conception was vaguely considered by some to be an event where future life took place in the womb (i.e., uterus). It was not known until the late 19th or early 20th century that human fertilization occurs in the oviduct.

The same objections the Church raised against mechanical or chemical contraception it raised against in vitro fertilization. There are many other aspects of the debate, and not all deal with natural law arguments. When eggs are produced after hormone stimulation, about 15 or so are removed by scraping the surface of the ovary. These eggs are then fertilized in a dish. An egg must have two polar bodies and two pronuclei to be considered potentially normal (not haploid, triploid, or polyploid). The fertilized gametes chosen for implantation all meet these criteria. If there are three or four such satisfactory preimplantation embryos, the rest of the potentially normal preimplantation embryos are stored and frozen in liquid nitrogen, in case the parents wish to try again should this round fail or if they later desire another child. The couple signs a legal document indicating how long these embryos may be stored, whether they can be used for research, whether they can be donated to an infertile couple, or whether they should be destroyed.

The destruction of fertilized eggs at any stage, pre- or postimplantation, is considered by some Catholics, some Protestant denominations (especially among Evangelical Christians), and some Orthodox Jewish groups to be an act of killing in violation of the biblical commandment "thou shalt not kill." The same argument about killing applies to the use of the inner mass cells of the blastocyst. These cells, as we noted in Chapter 14, are now called embryonic stem cells, and they have the potential for medical use to treat a variety of tissue-loss diseases.

The use of ICSI presents an unusual situation where genes that were once self-eliminating, especially sterility genes on the Y chromosome, are deliberately introduced to produce a son (if the Y chromosome is in that sperm) that will pass on the defect. Most single-gene male infertility is associated with Y chromosome mutations. An argument frequently made by the informed parents is that a disease ceases to be a disease if there is a treatment, and they do not see this as morally wrong. They see it as no worse than passing on genes for being nearsighted. Some geneticists do see a difference, because it increases the percentage of sterility every generation if natural elimination is subverted by medical science. This is not an easy issue to resolve, and there are many options now and in the not too distant future. Such parents could choose to have daughters instead of sons and thus not pass on Y-linked sterility genes. Such parents could elect (if it ever comes to pass) gene replacement by inserting the normal form of the gene into the defective stem cell line leading to sperm formation. Males with the defective Y chromosome gene could marry later or have fewer children than the average and thus delay the accumulation of mutations in the population. By having fewer children, many more sterile lines will die out by having only one daughter and no son.

The moral issues in these approaches are all based on utilitarian values. The greater good is having a child of one's own, even if that child is sterile. The religious objection to such behavior is not that the utilitarian values are wrong, but that the process involves sacrificing one or more fertilized eggs to allow such a child to come into existence. No doubt all the arguments would disappear except for natural law arguments of the Catholic Church if only one embryo is formed and used and works to produce an IVF child. This is not likely to happen because of the odds against any normal fertilization by ordinary intercourse causing a successful pregnancy.

All new fields of scientific knowledge, when applied, generate ethical and moral arguments about their applications. The field of assisted reproduction is rife with such issues. It is not just religious opposition to the medical intrusion into the field of human sexuality and reproduction; it is the legal limbo of deciding cases over ownership of fertilized eggs, the rights of divorced spouses to request destruction or preservation of fertilized eggs, the right of a paid woman who is carrying a child, neither of whose gametes are hers, to claim the child she bears because she has developed a maternal attachment to it. These cases fall into areas of contract law, divorce law, and child support laws and properly are handled by the lengthy process of judicial review as one test case after another is weighed against past judicial opinion. Some critics of assisted reproduction claim that these cases would not trouble our courts if the field would just cease to exist. They see the field as intrinsically immoral. Most infertile couples do not, and most physicians do not. In all likelihood, the desires of the infertile and the medical profession will prevail over those who object to the unnatural ways we address biological misfortune.

The costs of assisted reproduction are high (usually several thousand dollars) for in vitro fertilization and much higher if procedures involving donors are also used. The cost of certified sperm (free of HIV and other infectious germs and free of likely birth defects) is high, but it is even higher for certified eggs from a donor. Some cut their costs if they are not insured by allowing unused eggs or fertilized eggs to be used by other infertile couples. But all medical procedures are costly in the U.S., where there is virtually no regulation of the costs of medicine in a fee-for-service climate. Some critics would prefer a reproductive policy similar to that in effect for blood donations. It should be considered a gift relationship rather than a means for the desperate to earn income from use of their gametes or wombs. This might work ideally, but with nearly 10% of humanity infertile, it is not likely to attract enough people of good will to produce the needs of the many. In an imperfect world, imperfect compromises are the rule.

Also at issue is the use, especially if a person pays for it, of a surrogate womb (a woman hired to bear a child she will deliver but for which she has no biological claim). Those who object to this on grounds other than natural law violation may do so by invoking exploitation of the vulnerable. The issue is complicated because an unrelated woman is usually paid to carry the pregnancy that the infertile woman cannot because of some uterine defect. The surrogate is frequently someone who needs the money. Many critics of surrogate mothering claim that this is exploitation of the poor or poorly informed and that such women put themselves at risk both physically and mentally for doing this. There are fewer objections when a relative (mother or sister or aunt of the infertile woman) does this without being paid. Even when the exploitation argument is blunted by using such a close relative, there are concerns that this could lead to rivalry for the child by a sister or cousin or that this endangers the health of an aunt or mother of the sterile woman. In all cases of moral dispute, it is important to inform all the parties of what they are doing and what possible risks are involved, psychologically, legally, and physically. People do have to make moral choices when they decide whether to reproduce or remain a childless couple, whether they adopt or not, whether they use high-tech medicine or not. For this reason, most couples inquiring about their infertility problem receive many hours of counseling from the staff of the reproductive services office they are visiting. Ultimately, it is the decision by the users of these techniques to make that judgment. Making moral choices does not mean that those choices will please everyone.

The Argument on Not Playing God

One rarely hears serious arguments from philosophers and theologians about not playing God. It is more of a pop cultural belief. A major reason for this is that there is no agreement on what constitutes playing God in an inappropriate way. Who plays God? If you think about it, virtually everyone. To be a judge and pass a life sentence is playing God. To be a government leader and order a nation into war is playing God. To be a teacher and give students an F is playing God. To be a business executive and close down a factory is playing God. Every physician plays God when withholding or providing information. So does every parent in deciding what children should eat, wear, or say in public or at home.

What people mean when they say scientists should not play God is that they fear the unknown. They fear bad outcomes that might be of horrific damage to humanity or that will drastically change their conception of what it means to be human. The best way to prevent such imagined or possible horrors is to regulate science. We know that other institutions have failed us in the past. Governments have led countries astray, and we regulate that by revolution (think of 1776) or by the ballot box. Businesses have failed investors (think of 1929), and we did not abolish business—we created a Securities and Exchange Commission. This does not mean science is exempt from the same responsibilities to do no harm that other institutions, by law, are required to fulfill. It does mean that each new scientific field that is applied to human use must pass through scrutiny from agencies responsive to those voicing concerns. For assisted reproduction, there is ample evidence that the field has proceeded slowly and carefully, following guidelines for new medical procedures that apply to surgery and prescription medication.

A key concept in the idea of playing God is that we are intruding into an area where humans should not go. This is the Frankenstein sense of not playing God. It is invoked for almost all new technologies from new science. Why do we fear the new? It can change the way we deal with the world. For those white people before the Civil Rights movement, the idea of having blacks in movie theaters, restaurants, churches, or their own neighborhood was frightening. It was unnatural because they had learned to live in an era of white supremacy, and giving that up was not easy. Today we find no such shudders of fear if we see white and black soldiers together marching, eating, or fighting an enemy. The use of assisted reproduction made its first practitioners seem suspect (if not diabolical) to many critics of what they called test tube babies. A generation later, the objections are largely confined, not to lurid images of such babies suspended in tubes, but to more theoretical (usually religious) beliefs that the process is tainted by the murder of innocent zygotes or preimplantation embryos. Playing God arguments tend to fade faster than the religious objections to new reproductive technologies.

The Moral Response to Objections about Assisted Reproduction

For the nearly 10% of humanity facing sterility, having a child is an important part of their marriage or the partnership of living together. In well-developed countries, there are simply not enough adoptable children (especially newborn babies who are free of birth defects) to fulfill the requests from the sterile. State laws also vary on adoption policy, and there may be religious and racial limits on adoption. Physicians look upon infertility as a medical problem requiring medical procedures. The use of in vitro fertilization seems as appropriate to the practitioners in the field as the use of organ donation, blood transfusion, or surgery in the treatment of defects and diseases of other organ systems. The IVF practitioners claim they are not interested in eugenics, nor are their patients. They are there to provide a medical treatment for infertility.

Evidently, the public has agreed that infertility is a disease or medical disorder and not a judgment of fate that must be endured. The use of assisted reproduction is widespread, and several hundred thousand persons owe their lives to this procedure that is now a standard part of medicine around the world. Despite the objections by the Church, Roman Catholics are as likely to use IVF procedures as non-Catholics. The controversy over IVF as a Brave New World menace has disappeared. The Right-to-Life movement quietly dropped its objections to IVF technology when it threatened to split the anti-abortion movement. No doubt there is unease among both Catholics and those Protestants and Jews who feel uncomfortable with some of the IVF procedures and consequences of embryo storage in liquid nitrogen, but they have not made it a moral crusade of the same intensity as elective abortions. There is also little doubt that users of assisted reproduction invoke utilitarian ethics as they choose their options in response to the reality of being infertile.

Notes and References

1. The most useful source for current views and activities in the field of assisted reproduction is a journal, *Fertility and Sterility*, which has research and clinical articles on all aspects of human infertility and its treatment. What is notable about this journal is its inclusion in each issue of an ethical discussion of new techniques and evaluations of techniques that have been tried for several years.
2. My wife Nedra for 13 years was an IVF embryologist. She helped put together 3000 live-born children (she says she made a village), and during those years the rate of success for a live-born child moved from 10% to 40%. Much of that success was an outcome of the use of more advanced preimplantation embryos.
3. The Y chromosome is unusual because it has few genes other than those that help male sex determination in the embryo and genes for fertility associated with sperm production. Factors for female development and egg development are associated with both the X and the autosomes. Thus, use of defective sperm is almost always associated with Y-chromosome abnormalities that will be passed on by ICSI. In females, both the X and the autosomal gene defects leading to infertility will probably be heterozygous and not expressed through the child but will be expressed at some future generation.
4. Virtually no one uses that term now to describe IVF babies. It illustrates how the initial reaction of the press and the public relies on myths and stereotypes and later assimilates the process as a technical or clinical one in the terminology of the new medical field. The first baby born by IVF came into being on July 25, 1978, in Oldham, England, and she was named Louise Joy Brown. The first American IVF baby came into being in 1981, and she was given the name Elizabeth Jordan Carr. The physicians who supervised her conception and birth were Howard and Georgeanna Jones, of the Jones Institute in Norfolk, Virginia, where Edwards had trained.
5. Robert Edwards and Patrick Steptoe, *Matter of Life: The Story of A Medical Breakthrough* (Morrow, New York, 1980). The authors describe their career paths that led them to collaborate on this historic birth.

PART 7

ASSESSING BAD OUTCOMES

S CIENTISTS ARE LESS LIKELY THAN THEOLOGIANS or philosophers to assess in detail the benefits and harms that may stem from their work, especially as it is applied to humanity. With rare exceptions, all such applications can be described as having good intentions. The good intentions may not be for the subjects involved, as Nazi biological sciences showed, but instead for some higher good by those of a particular ideology. The Holocaust and the American eugenics movement well illustrate that distorted idealism. In the first chapter of this section, I try to illustrate these concerns and the responses to them by citing the criminal justice system as one model for determining whether a crime has been committed or whether civil lawsuits are the appropriate responses to bad outcomes. The spectrum of possible indictments and the spectrum of awards considered and granted by juries in civil lawsuits reflect a somewhat quantitative model in the minds of those involved in the legal system. I would not be surprised to find a similar quantitative scheme developed by scientists in assessing what should be done to punish those who have harmed others through their science. It is less important to have a numerical scale than to have something like the legal system, where severity is usually proportional to the punishment and fines are meted out to those found guilty.

In the second chapter of this section, I discuss the way in which we can use history and the liberal arts to assess why we do what we do. Motivations are usually more complex or contradictory than we wish them to be. We lack an effective science of how our actions and beliefs come about, and it is virtually impossible to predict human behavior, whether it is explaining Hitler, the psychotic activity of those perpetrating a massacre in a public school, or the capacity of respectable professionals to participate in state-sponsored killing. I cannot claim to have answers, but I do make a plea that it is a valid field for research.

In the third of these chapters, I take on what is unpleasant for most scientists to confront. Scientists are motivated to detect new knowledge by studying some aspect of the material world for which they have both the skills and the curiosity to explore. When they do so, their findings may be at odds with popular prejudice (e.g., the alleged inferiority of Jews to Aryans or blacks to whites), with the paranormal (e.g., the belief that some people have the ability to see future events or to move or alter the shape of objects by looking at them), or with a particular religious tradition (e.g., the belief that the sequence of creation events in *Genesis* is a literally true account of an historical event). Scientific findings are often at odds with such prevailing beliefs, but I illustrate that in the long run, science prevails, and most religions accommodate the new knowledge without suffering a collapse of faith (as I illustrate for the centuries following the trial of Galileo for heresy).

Is it important to discuss these conflicts between the worldview of science and the worldview of those relying on unexamined bias, ignorance, or a narrow interpretation of theological views? I argue that it is essential for science to defend its findings and to expand its efforts to communicate those findings to a public that may be largely ignorant of both how science works and what its contributions are.

In the final chapter of this section, I explore the present concerns of science as it enters the 21st century. There may be setbacks for science imposed by government action and a long tension between science and its critics. In the long run, the realization of the good intentions of science and an increased appreciation of its worldview will be far more likely than these temporary setbacks.

16

Quantifying Evil or Bad Outcomes

SCIENTISTS DO NOT LIKE TO USE TERMS that are subjective, difficult to define, or lacking a quantitative basis. For this reason, the term "evil" (as a noun or an adjective) seems too vague and judgmental to be of use in evaluating the bad outcomes of science.[1] However, such a quantitative system, I believe, already exists, and it is to be found in our courts and law books. We have a recognized spectrum for killing including premeditated murder, killing with disregard for life, manslaughter, killing while under the influence of alcohol, accidental killing, killing in self-defense, killing in wartime engagement, and killing by reason of insanity. I'm sure there are other deaths that could be slotted into this spectrum. The same effort would apply to assaults and to loss of property. We also have criminal prosecutions, and we distinguish those bad outcomes from civil actions resulting in financial compensation.

This is not a perfect system. Some would like to insert their religious value system into a secular legal system, and in times past, that did occur. However, today in secular societies, blasphemy is no longer a crime, an atheist does not face a death sentence, and a woman who asks for an abortion is not prosecuted for attempted murder. There are countries today where such theocratic rule does exist and an adulterous woman (but not usually a male) can be stoned to death. Fortunately, most industrial countries are secular, and they have settled for a spectrum of evil or bad outcomes that is purged of religious crimes. In fact, the term evil does not have to appear at all in our secular system of assessing bad outcomes, although both prosecutors and judges worked up by a particularly heinous crime may use that term in demanding justice or passing judgment.

Just as we distinguish between criminal and civil law, I believe we could reasonably classify the bad outcomes of science into five categories:

1. Criminal acts that involve behavior done knowingly which result in death, injury, torture, or other psychological and financial harm, done by those whose values were antisocial, banal, and self-serving. This would include many top Nazi war criminals prosecuted in Nuremberg trials, Lysenko and those of his supporters who suppressed geneticists, and scientists who manipulated experimental results leading to harm in those who believed the work was done thoroughly and consistently (as in the thalidomide case).

2. Acts leading to harm from those who advocate social experiments with inadequate evidence to justify their proposals. Laughlin's efforts as a lobbyist for eugenic laws, Sharp's campaign for laws for compulsory sterilization, and the race hygiene ideology promoted by Baur, Fischer, and Lenz in their human genetics textbook would be such examples.

3. Nonmalicious acts which involve bad outcomes that should be compensated by civil law more often than by criminal law. This would include malpractice in medicine, or engineers who were careless and, without malicious intent, violated their own safety procedures in Three Mile Island and Chernobyl.

4. Acts of plagiarism or using falsified data claiming scientific findings or interpretations that mislead readers and set back research or understanding. These rarely result in criminal prosecution and are usually handled by resignation from an institution or by firing the perpetrator of the fraud and by public retraction of such work in the published scientific literature.

5. Noncriminal acts that have violated no criminal or civil law but which merit disapproval by the press, the public, and the judgment of history. I would include in this category those who looked the other way when they had an opportunity to protest wrongdoing by their colleagues or superiors.

We would have much more divided opinion on a sixth category of bad outcomes, involving participation in the design and construction of weapons of mass destruction. One cannot rationalize away the fact that bad outcomes are part of the consequence of using these weapons, because civilians are a major target of them (e.g., germ warfare, atomic bombs, poison gas). The problem lies in the motivations of those who participate in their design and construction. It makes no difference whether the persons work for Nazis, Communist Party politburos, or democracies. All say the same thing—they do it for patriotic reasons, including survival, national security, or defensive response to a threat. In every case, utilitarian ethics is used to justify a higher good for designing or even using the weapons than for refusing to work on such weapons. Also used is an older eye-for-an-eye attitude—if we don't work on this, our enemies will.[2] This tit-for-tat outlook does not apply, however, to our response to terrorism. Democracies do not promote terrorizing civilians in the home countries of terrorists in response to acts of terror. They do not set up their own programs for suicide bombers, beheading kidnapped civilians, or blowing up schools with children in them. From a strictly utilitarian perspective, however, the deaths from terrorism are minor (so far) compared to the deaths from the use of weapons of mass destruction (compare the number of civilian Hiroshima and Nagasaki deaths to those in the World Trade Center or the Pentagon). Despite that inconsistency in utilitarian ethics, we are unable to pass a negative judgment on those in our own country who have worked on or used weapons of mass destruction to kill civilians. Why is this so? I believe it is because we depend on our constellation of values, inconsistent as they are, to guide us. I also believe that in times of peril, patriotism trumps our personal values, and we fall back on the tainted rationalization that the end justifies the means.

Although the issue of guilt or rationalized justification still lingers in the minds of participants whose work leads to bad outcomes, there are many scientists who should feel no guilt at all because they are or were falsely accused of bad outcomes that have not happened. This was true for those participating in the first years of recombinant DNA technology when concerned scientists, responding to a few scientists who worried about possible bad outcomes in a yet-to-be-launched field of applied and basic science, suggested a conference to hash out those issues. Those famous Asilomar conferences on recombinant DNA technology led to a very responsible suggestion—scientists would regulate their work and set up criteria on containment facilities and a hierarchy of risks that could be altered as each lower rung of possible risk was demonstrated to be risk-free. Those guidelines have worked, and in a generation of recombinant DNA technology, no runaway infections have occurred through accidental introduction of pathogenic genes from one species into another. This does not mean that such genes cannot be inserted deliberately (as in germ warfare laboratories), but it does mean that safeguards to prevent accidental chimeric cells with deadly consequences are so effective that no one raises that issue today for pharmaceutical companies using recombinant DNA technology to produce hormones and other medically useful products.

It is also important to draw a distinction between basic science, where the quest is adding to our knowledge of the universe, and applied science, where some good intention motivates those doing that science. The presence of good intentions does not justify the applied science unless some forethought goes into the uses and development of the applied science. It might be a good idea to build a *Titanic*, a Tacoma Narrows Bridge, or a Chernobyl reactor, but that does not exempt the scientists from lawsuits or prosecution for flawed designs or operations. It may be a good idea to have a cheap, nonaddictive sedative, but marketing a product like thalidomide that fails two of its three tests as a sedative suggests a less than full disclosure on its effectiveness. Even more disturbing for the thalidomide case is the deception practiced in misleading those physicians who wrote in about complaints of peripheral neuritis and constipation from its use. Today, drug companies are required by law (and the potential threat of lawsuits) to list such reported side effects. Those are instances where there is a lot of agreement that regulation works. No such regulation exists, however, for the design of military weapons that can kill the innocent, and at present, we can only appeal to the individual consciences of scientists to reflect on their motivations and values.

These bad outcomes are rarely part of those who do basic research. Learning how the universe works often does provide a science that can be applied, but the motivations for basic science stress the pleasure of new knowledge in our comprehension of the universe. True, a few basic scientists may also see useful applications (or even debatable ones, as Lise Meitner's discovery quickly suggested to physicists around the world that atomic energy could be used for commercial and military applications).[3] That is true of all knowledge, but Gregor Mendel did not work on hybridization with peas to apply his findings to germ warfare or the sterilization of the unfit. He could not possibly have imagined those abuses of knowledge, and we cannot hold basic scientists responsible for the abuses of their knowledge. We could, and should, hold scientists responsible for their suggestions on how to apply new knowledge—a good case would be David Starr Jordan's classification of the Tribe of Ishmael in Indiana as a human weed ("like devil grass") or as a form of parasitism (because of their dependence on public assistance) and as a justification for isolating their children from their parents to prevent another generation trained in social failure.[4]

Some scientists have proposed a scientific equivalent to the Hippocratic Oath—to do no harm through science and to apply new knowledge with ethical reflection. That is unlikely to happen, because the route between science and its bad outcomes is often indirect, requiring the participation of other parties such as the military or management in a corporation for the way products are promoted or used. In medicine, the doctor–patient relationship is much more direct, and the oath is a constant reminder of that closeness. I would also hesitate to impose, but would not necessarily oppose, mandatory courses in ethics for science students, because there is a danger of indoctrination. Both the USSR and Nazi Germany believed very much in the social responsibility of the scientist to the state's needs. It may be more important for a scientist to have a liberal arts undergraduate education that explores ethics and values in elective courses on the history of science, the sociology of science, and formal philosophy courses than to tack on a graduate-level short course that strikes the student as propaganda or contrived. For that reason, such approaches should be debated before implementation by scientists and their colleagues who would be teaching such courses.

Notes and References

1. I was surprised to see how many scientists in a Chautauqua course I taught on this theme of good intentions and bad outcomes felt that the use of the term evil, even for Nazi physicians tried for war crimes, was inappropriate. They did not argue against their being tried, they argued that being a criminal and being evil are different issues. By calling a person or a person's acts evil, they claimed, one changes the objectivity of science and fails to take recognition of the constructed nature of crime. They believed that if the Nazis had won the war, there would have been no concept of war crimes for what they did.
2. This rationalization is widely used along with the equally dubious argument that arms, war, retaliation, and other tit-for-tat reciprocity are part of human nature. Human nature is more often perceived in negative impulses than in gregarious ones.
3. Einstein's famous letter of 1939 to FDR warning of the possibility of an atomic bomb is considered a positive example of this: Facsimile text of the letter can be seen at http://www.mbe.doe.gov/me70/manhattan/einstein_letter_photograph.htm/.
4. See David Starr Jordan, *Footnotes to Evolution: A Series of Popular Addresses on the Evolution of Life* (Appleton, New York, 1899). See Chapter 11, "Degeneration" and Chapter 12, "Hereditary inefficiency."

17

Science, History, and Responsibility

A MISSING COMPONENT IN DEBATES about abuses of science is history. By going back to primary sources and seeing the connections among motivations, values, and traditions prevalent at the time a work of science was developed or applied, we can gain insights into why things went wrong without projecting spurious motivations and moral views of the present. Those of a newer generation are more apt to believe or to construct a myth about the past. I do not say this to absolve wrongdoing or responsibility for bad outcomes. Rather, I believe we have a better chance of preventing future mistakes by knowing what actually happened and why those involved in scientific bad outcomes did what they did.

Lessons from the Use of Theory for Social Policy

When Harry Clay Sharp read Albert Ochsner's article on vasectomy for degenerates in 1899 in the *Journal of the American Medical Association*, he was primed to respect the view of a senior medical scientist he admired. Ochsner was a founder of the American College of Surgeons and had published many books and articles on surgical procedures as well as on the management and organization of hospitals.[1] He had studied medicine with Rudolph Virchow, one of the major figures in 19th-century medicine. Ochsner, who lived and practiced in Chicago, was not a zealot, and among his voluminous publications, this was his only article on degeneracy. It was an observation or reflection he made that added to an ongoing debate in American society. Degeneracy in the 1890s was believed to cause psychosis,

193

mental retardation, vagrancy, criminality, and pauperism. Popular journals of that time, as well as occasional editorials in medical journals, raised concern about the alleged degeneracy of a portion of American society. It was placing a burden on the middle-class public who had to pay taxes to support prisons, asylums, public hospitals, and welfare or charity programs for the degenerates. That same concern was being voiced in Great Britain, France, and Germany among the leading industrial nations. Degeneracy was believed to be real and very likely caused by bad environments. What differed in 1899 from thinking a century later was that most scientists then believed a bad environment altered heredity, and the altered heredity was passed on. Some of the debate concerned whether good environments could reverse such degeneracy. By 1899, sentiment was going against a simple Lamarckian plasticity of heredity and favoring August Weismann's views that the alleged debased heredity was permanently fixed in the germ plasm.[2]

We can analyze Ochsner's recommendation to use surgery, preferably vasectomy, to prevent degeneracy from being transmitted:

- The technique has no negative physiological consequence other than sterilization, unlike castration, which drastically changes a male's body and mind.

- About 10% of humanity was classified by sociologists as degenerate in the 1890s (they were often called the "submerged tenth.")

- Ochsner believed, as did his contemporaries, that degeneracy was transmitted in a like-to-like manner: Degenerates produce degenerates; normal people produce normal offspring.

- Ochsner believed it is a government's responsibility to keep its population healthy by sponsoring effective public health.

- Ochsner believed that public health included preventive medicine, including the sterilization of the unfit.[3]

Although this motivational listing may have seemed logical to Ochsner and the medical profession in the 1890s, it has some unexamined assumptions:

- Ochsner had no evidence that degeneracy was inherited. His acceptance of that view reflects the poor quality of biomedical science in his time.

- Ochsner assumed a society can do what it wants to members of its society in order to benefit the entire population. Respect for the autonomy of the individual required the movements of the 1960s, which in the U.S. permeated from civil rights, to women's rights, and to patients' rights.

- Ochsner assumed a role for medicine making the physician judge, jury, and executioner of medical decisions about who are the degenerates and what should be done to them. It is part of the paternalistic tradition in medicine that prevailed largely until the 1970s.

Ochsner's failing, in my view, was his lack of reflection on the proposition he was asking society to initiate. He had shifted a social and economic problem to a medical problem without rigid evidence to justify his request.[4] That is arrogance, and it implies a disdain for the views of other social reformers who believed that the plight of the degenerate was not a reflection of bad germ plasm, but of a selfish society that does not want to face the costs in time and money to do effective reforms. A major difference between Ochsner and Sharp is that Sharp carried out Ochsner's proposal first by stealth (allegedly treating his patients for onanism by performing vasectomies on them) and then by design (campaigning and succeeding in getting the state legislature and governor to pass a compulsory sterilization law). Had there been a Nuremberg trial parallel, Ochsner would have been like Verschuer, Lenz, or Fischer, and nothing would have happened to his career; but Sharp, for executing the ideas of Ochsner, would have faced arrest and sentencing for the 500 or more prisoners he vasectomized. Then as now, people who make suggestions leading to bad outcomes are less liable to criminal prosecution than are those who carry out those suggestions. That legal challenge never happened to Ochsner or Sharp because the American government (state and federal) endorsed their views through legislation or court decisions. The Nazis were prosecuted because they lost the war, and a different system of justice rejected a Nazi policy that was secretly ordered by Hitler (there was no formal German law allowing the Holocaust or the medical experimentations on subjects without their consent).

We cannot rewrite history or change what has already happened. We can learn from this historical episode that scientists with well-intentioned ideas applied to social policy need to take care. They need to be reflective. They need to test their ideas with critics before they offer them to the public through publications. If they do publish first, it is important for other scientists who anticipate bad outcomes from those ideas to speak out. I don't favor pie-in-the-face confrontations or disruptions of scientific meetings by those who may be as zealous (or wrong) as those they dislike. What is persuasive to scientists are evidence and reasoned arguments, not threats or intimidation.

Lessons from Scientists Who Apply Their Knowledge

The application of science necessarily creates a moral climate different from the objectivity of basic science. In basic science, integrity is the chief ethical value. The study is done honestly, and the motivation is the finding of new knowledge. There are occasions when that objectivity of basic science can be stressed. I think of Alfred Kinsey at Indiana University in the late 1940s and early 1950s when he was attacked for studying human sexual behavior by using scientific approaches.[5] I was a graduate student when I first met him, and I was one of the last persons to see him alive. I happened to come in early on a Saturday morning, and I encountered him gasping for breath as he was walking up the steps of Jordan Hall to his office. He had just sat down on the steps. I asked him if he needed help. He said he'd be fine and only had a few more steps to go. I asked him why he didn't take the elevator. He said it was too slow and he had to prepare for a group of people who had signed up to learn about his Institute for Sex Research. I thought that was remarkable dedication—an international scholar who was so committed to his science that he would drag himself to work and not delegate that activity to a staff member. He died shortly after that encounter. Kinsey was fortunate that the university president Herman Wells had a strong commitment to academic freedom. Wells defied members of the board of trustees who wanted Kinsey ousted. He explained to them that science cannot work if it is intimidated and that Indiana University would fail as an institution if it excluded human sexuality as a subject for research by a scientist with proven scholarly habits.

Applied research is inherently compromised. There is not only an obligation to carry over the values of research objectivity, but also an obligation of loyalty to the employer or institution. Some employers demand secrecy. Some companies or institutions may delay or forbid publication of findings. The choice of projects, compared to a university, may be greatly restricted. If scientists believe they are being seriously compromised, they can quit or look for work elsewhere. That is not easy for scientists who have families to support or who are well established in a community.

Quite different is a poorly explored influence of the social thinking of the day on a scientist. A very gifted scientist, Edward Murray East, a major contributor to our knowledge of the development of hybrid corn and to our understanding of how hybrid vigor arises, was a Harvard professor who clearly enjoyed his celebrity. His popular book *Inbreeding and Outbreeding* was a scientific best-seller. But in his zeal to popularize the new science of genetics and its benefits to society, he was also a racist in his eugenic views.[6] He looked upon blacks as inferior and depicted their personalities with the same unexamined bigotry as did early 18th- and 19th-century scientists such as Linnaeus or de Gobineau. East was not alone in that racism of the 1920s. It was widespread in American culture, from racist jokes to racist stereotyping with legal restrictions on the immigration of Orientals (and later all others thought to be undesirable) through the 1921 and 1924 Johnson Acts that established a quota system for admissions to the U.S. based on the 1890 census.

There was an almost universal acceptance among whites that blacks should not aspire to success in more than the ministry, entertainment, or unskilled labor. What is difficult to interpret is the double standard many scientists applied to their own work on animals and plants and to their work on humans. They used their critical skills to design controlled experiments; they demanded proven facts; and they applied other objective approaches to maize, fruit flies, or mice. Yet they made sweeping conclusions about human behavior based on very broad theories which often lacked supporting evidence or careful analysis. Why was this so? There is some suspension of careful reason and the scientific tradition when blacks are described as "happy-go-lucky," when Jews are described as parasites and morally depraved, when Italians are depicted as drawn to organized crime, and when Slavs and Serbs are depicted as having an excess of feebleminded children. Unfortunately, these were not only sentiments of bigots; they were also the sentiments of many geneticists and scholars of repute in the eugenics movement of the first half of the 20th century. Such sentiments suggest that not only can patriotism corrupt values and make people do things that are harmful to others, but also social beliefs that are pervasive can eclipse the scientist's normal tendency for objectivity.

Add to that the role of religion in shaping views at variance with scientific standards, from Galileo to the present. There are a small number of contemporary scientists whose religious training and faith prevent them from accepting a scientific worldview at odds with their religious traditions, especially for issues on the origins of components of the universe, the age of stars and galaxies, the origin of life on earth, or the age of fossil remains. It is one thing to reject a theory of evolution by natural selection on religious grounds, but it is more difficult to reject carbon-14 dating, isotopic uranium dating, red-shifts as measures of the age of light arriving on earth, age determination by counting sedimentary layers of mineral deposits, and tree rings in ancient wood, all of which indicate a much longer duration of life on earth and greater age of the earth and the solar system than many Evangelical Christians (as well as many Orthodox Jews and Moslems) would have us believe because of their adherence to a young earth model favored by a few theologians.

The Divided Self for the Scientist

In many ways, the applied scientist faces the same problems (but not usually in so traumatic a form) that were depicted in Greek tragedy. In Sophocles' play *Antigone*, the lead character Antigone has what literary critics call a "divided self."[7] She is loyal to her king (her uncle Creon) but she is also loyal to her dead brother, Polynices, lying on the battlefield after being slain by Creon's army for his insurrection. As punishment for that betrayal of the state, Creon orders Polynices' body to remain there, to decompose or to be left for scavengers to eat. Antigone is divided between loyalties to her king or to her brother's memory; she chooses to defy Creon and bury her brother. Oppenheimer faced the divided self in his own work at Los Alamos. At first he had unwavering loyalty to General Leslie Groves, never disputing the use of the bomb to end the war against Japan. But as he saw the magnitude of destruction (far greater than he had imagined) and reflected on what a nuclear military might bring into being, he had his doubts and began a campaign to oppose work on an even greater weapon prepared for future wars—the hydrogen bomb. Patriotism worked while there was a war, and Oppenheimer went with "sin," as he later portrayed it. After the war, that patriotism became doubtful and the possibility of millions of deaths from future atomic wars outweighed, in Oppenheimer's mind, the loyalty demanded for unquestioned service to the state. There will always be citizens who put their country first, even if it compromises their otherwise impeccable values. It is a dangerous loyalty, as many a Nazi found out when the war ended and their crimes against humanity replaced their belief that they were being loyal to a state purifying itself of noxious degenerate components.

Most applied scientists do not face such tormenting decisions about their role, and most institutions and companies are guided by ethical principles. The rogue cases described in this book are exceptions, but their bad outcomes have had very serious consequences for society. It is worth listing some of these abuses of mutual trust:

- It is wrong to withhold information about harm to the health of those who manufacture a harmful substance or who are exposed to such a product. Lawsuits abound in the asbestos industry, pesticide and herbicide industries, and other industries where toxic products are made. Care must be taken that such products do not contaminate those who work in the facility or in the community where such products are used and wastes are produced. Neither the urgency of war nor the prospect of financial losses justifies doing harm to the health of others. The solution to that potential harm is to provide healthier work environments and better disposal of toxic products and to include those costs in the product's sale price.

- It is wrong to deceive the public about the safety or efficacy of a drug they buy in good faith. We can understand some past bad outcomes (such as DES) as having arisen from a low level of quality of medical research that claimed DES prevented miscarriages, but the lesson learned from DES (as in thalidomide) is better regulation of the drug industry, even if it delays the introduction of new drugs into the market. Any medication prescribed for use (or likely to be used unknowingly) during the first trimester of pregnancy should be carefully regulated.

- It is wrong to despoil the environment with wastes or products that do serious harm to the ecosystem. We are often indifferent to the stewardship we are obliged to show to the rest of the living world. In the 1950s and 1960s, some of these blighted environments forced public response. Killer smogs in London led to many deaths among the elderly, asthmatics, and others with pulmonary problems. I remember such smogs in New York City in the 1940s that smarted one's eyes and made breathing difficult. I remember seeing during the early 1960s the dead trees along the sides of the mountains surrounding the valley in which Los Angeles was situated. We still remember the health problems imposed on people living near toxic waste dumping grounds such as the Love Canal in upstate New York. We also remember the power of Rachel Carson's *Silent Spring* that launched the environmental movement. Working out a balance between human needs and the health of the environment is never easy and never perfect, but writing off the environment as a romantic frill is dangerous to our health and the quality of life we need.

- It is wrong to assign attributes of human behavior, especially those that demean one population or ethnic group as inferior to another, without the same standards of scientific rigor that apply to studies of animals in laboratory settings. Those who do so may not consider themselves racist, and it is better to challenge such claims on the basis of their flawed research (as is likely the case) than by assuming their motivation is simple bigotry.

This is an assessment based on experience and the consensus of society that it wishes to be protected from the abuses of those who are selfish, indifferent, self-deceived, or naive. We can get that protection by regulation, by legislation, from the courts, or by the conscientious application of our own values to the scientific work we do.

Why Do We Do What We Do?

The question that cannot be answered without dispute is why we do what we do. I have argued that most scientists, like most people, have good intentions, and occasionally they lead to bad outcomes. We have looked at some of the ethical positions taken by participants in these bad outcomes. Is there a way of doing a more systematic analysis of why things go wrong? We can certainly list the attempts so we can make some evaluation of their strengths and hope we or future scholars can point the way to where research should be done to get better insights. Excluding religious concepts such as original sin, which are based on faith, religious teachings, or biblical authority, the major secular interpretations of what can go wrong include:

- Patriotism as a virtue or obligation is built in for most of humanity as effectively as a belief in the virtue of one's religion or one's family. It is the social glue that binds diverse people, unrelated to one another, into a common identity. Failure to support the policies of one's government during times of crisis or war is almost certain to lead to ostracism; charges of disloyalty, treason, or cowardice; or other negative reaction. Whatever their personal values, most people are willing to obey rather than resist the allegedly good intentions of their government. I know of no scientific studies of the way patriotism causes people to suspend their otherwise socially approved values that prevail in peacetime.[8]

- Freud proposed a triune mind consisting of id, ego, and superego. The id is the relatively undisciplined, pleasure-seeking, and selfish component of the mind. It is shut down or kept in check by the superego. The ego is the operating system in day-to-day conduct, with the superego and the id shoved aside so the business of living can go on. Freud was the first to interpret war as a consequence of the breakdown of the superego in his

1930 classic, *Civilization and Its Discontents*.[9] Other psychiatrists, with less reliance on Freudian interpretations, have also explored this tension between the individual and the state. Erich Fromm's *Escape from Freedom*[10] is an excellent example. Most scientists dismiss Freud (or Fromm) as offering stories or myths without scientific evidence and see little difference between Freud's triune mind and religious concepts such as original sin.

- I learned about a very different psychiatric approach from Maurice Walsh, who had been a psychiatrist in the Army during World War II. He had evaluated the Dominican dictator Trujillo, and he was assigned to evaluate Rudolph Hess (Hitler's early confidant and secretary of the Nazi party who parachuted into England demanding England's surrender). Walsh thought the totalitarian leaders he interviewed were psychotic, and he wondered why normal people listened to them when they ran for office but would walk away from them if they just heard them on a soapbox in a park. He believed there was a field of leader–follower relations that needed to be explored by psychiatry and psychology.[11]

- Evolutionary psychologists have built many models of aggressive and altruistic behavior, as well as strategy games for cheating or not cheating. Their models are based on evolutionary success (the capacity to live, reproduce, and raise a family so one can pass on one's genes). Most of the evolutionary models invoke kinship support and non-kin enmity (unless trade-offs or intimidation brings about communities).[12] So far, this school of thought lacks the type of genetics most geneticists seek in experimental fields—genes that can be mapped and sequenced, whose products can be isolated, and whose components can be put to test in animal models or other experimental settings.

- I was struck by a remark made by the South African writer and Nobelist J. M. Coetzee in his second volume on his life (*Youth*).[13] Coetzee majored in mathematics and used that background to support himself when he left South Africa and went to England. He mentions working on an advanced military plane (that required his obtaining security clearance), not because he was patriotic, but because he wanted to see what evil felt like! Some people may be drawn to activities that harm others not because they are banal and want money, recognition, or power, but because they have no other outlet. Without the institutional sanction, harming others would be considered criminal or psychotic behavior. To some degree, Coetzee's personality can be described as Faustian. In Goethe's *Faust* (unlike the Faust most people know from Gounod's opera), Faust the scientist makes a pact with Mephistopheles to experience life in all its variety to see whether there is any aspect of it he wants to repeat. With Mephistopheles' help, he experiences love, power, war, the acquisition of wealth, dabbling in substance abuse and gambling, and delving into the occult. None of it does he find fulfilling, although he leaves a wake of bad outcomes behind him. His redemption comes at the end of Book II, when he realizes that by using science for human betterment—draining swamps to eliminate malaria, creating a harbor for world trade, building a city on the drained land, creating new arable land to feed the urban environment he helped create—he had found something worth repeating. In Goethe's *Faust*, Mephistopheles technically wins the bet, but Faust is claimed by God and ascends to heaven. In this model of behavior, some scientists would have Faustian personalities.

- Psychologists who have studied obedience (such as Stanley Milgram and P.G. Zimbardo) have shown that many college students have the capacity to do harmful things to others.[14] They support Arendt's and Lifton's thesis that ordinary people are capable of doing extraordinary harmful acts when they place their trust in authority. At the same time, there are some people who resist such authority and do not set their values on a back burner. No one knows the reasons why either group behaves the way it does.

I do not know whether any of these approaches will lead to a scientific understanding of why bad outcomes arise from good intentions. What puzzles me is that the problem is of importance to human survival and well-being, yet it is not a major field of research where scientists apply their more objective standards to gain an understanding. It should be.

Notes and References

1. A.J. Ochsner, "Surgical treatment of habitual criminals." *J. Am. Med. Assoc.* **32** (1899): 867–968.
2. See Elof Carlson, *The Unfit: A History of a Bad Idea* (Cold Spring Harbor Laboratory Press, Cold Spring Harbor, New York, 2001). Chapter 9, "Hereditary units and the pessimism of the germ plasm" provides an account of this conflict.
3. Anonymous. "Albert John Ochsner (1858–1925)." *J. Am. Med. Assoc.* **85** (1925): 374. This obituary reveals that after his MD from Rush Medical School in Chicago, Ochsner spent two years in Vienna and Berlin where he studied with Virchow and picked up an interest in public hygiene from him. A. Ploetz, who founded race hygiene, was also a student of Virchow. I do not know whether Virchow taught his students about degeneracy theory.

4. This shift is documented in historical detail and ably interpreted by Philip R. Reilly, *The surgical solution. A History of Involuntary Sterilization in the United States* (Johns Hopkins University Press, Baltimore, 1992).

5. James H. Jones, *Alfred C. Kinsey: A Public/Private Life* (W.W. Norton, New York, 1997). The author of *Bad Blood* examines some of the seamier side of Kinsey's life in trying to assess his motivations for studying human sexuality. Kinsey's homosexual liaisons and bisexuality were unknown to me when I was a student at Indiana University.

6. E.M. East, "Population." *Sci. Monthly* **10** (1920): 603–624.

7. I am grateful to Rose Zimbardo for providing that concept in her course on literature and the human life cycle that I attended in the Federated Learning Community of 1986 on the theme of "Social and Ethical Issues in the Life Sciences."

8. It may be that a thorough, objective, scholarly study of patriotism is taboo. Patriotism is so fundamental to national identity and the role of the state in our lives, that to question its demands or explore the bad outcomes it tolerates would ostracize anyone taking a look at it in a dispassionate way.

9. S. Freud, *Civilization and Its Discontents* (W.W. Norton, New York, 1930) (English version translated by James Strachey, 1961).

10. Erich Fromm, *Escape from Freedom* (Farrar and Rinehart, New York, paperback version, 1994) (Henry Holt, New York, 1941).

11. Maurice N. Walsh, *War and the Human Race* (Elsevier, New York, 1971).

12. Steven Pinker, *The Blank Slate: The Modern Denial of Human Nature* (Viking, New York, 2002); also, Paul Bingham, "Human uniqueness: A general theory." *Quart. Rev. Biol.* **74** (1999): 133–169.

13. J.M. Coetzee, *Youth: Scenes from a Provincial Life II* (Penguin, New York, 2001).

14. Stanley Milgram, "Behavioral study of obedience." *J. Abnorm. Soc. Psychol.* **67** (1963): 371–378; also S. Milgram, *Obedience to Authority: An Experimental View* (Harper and Row, New York, 1974); C. Haney, W.C. Banks, and P.G. Zimbardo, "Study of prisoners and guards in a simulated prison." *Naval Res. Rev.* **9** (1973): 1–17.

18

How Science Changes Our Worldview for the Better

SEVERAL YEARS AGO, I READ A NEW TRANSLATION by Robert Fagles of Homer's *Odyssey*, and I was struck by the worldview of Greeks who lived some 3000 or more years ago.[1] Everything was controlled by their gods. There were no natural phenomena. Aeolus blew the winds that filled the sails or slackened them. Several gods took turns guiding the arrows to their targets. Poseidon stirred up the seas into angry swells. Nothing was done by humanity without some assist from a pleased or offended deity. It was Helios who put the sun in a chariot and hauled it from dawn to dusk across the sky. It was Iris whose multicolored scarf formed the rainbow and colors of the sky. Thunder and lightning were formed on Vulcan's forge. To live in such an insecure universe, people were forced to make sacrifices to appropriate gods. A millennium later, Greek scholars were depicting a more rational world. They recognized that the earth is a sphere and even measured its circumference with surprising accuracy. Reading Aristotle is profoundly different from reading the *Odyssey*. Fatalism and a god-saturated reality are replaced by reason, and everything that happens in nature has a material cause.

When I read Dante as an undergraduate, I noted his perception of the material world. The earth was surrounded by spheres to which the planets, sun, and moon were attached. Everything moved around the earth at least in 24-hour cycles. A similar descending series of spheres existed within the earth as one moved through the layers of hell to the frigid core at its center. Galileo, we noted earlier, provided the first scientific evidence that this was a false view of the universe.[2] He used a telescope he had constructed himself and directed it at the nighttime sky.

He saw mountains and craters on the moon that cast shadows, demonstrating that the moon was a material body. He saw the four moons of Jupiter and named them for his Medici patron. He also followed their movements, showing that they moved around Jupiter, and he carefully computed their orbits and was able to predict where they would be on any future day. He noted that Venus had phases like the moon, and he inferred from this that it, too, must be a material body reflecting rather than generating its light to the earth. He saw sunspots on the sun, and from their reappearance, he could calculate that the sun rotates, and he determined that it rotated about every 24 days. Galileo was not speculating, as Copernicus was, that the sun was a solar system and a star with planets, including earth, revolving around it in near-circular orbits. Galileo gave convincing evidence, at least to his own satisfaction, that the solar system was a material universe and the material world was not limited to the earth. His findings dismayed many of his fellow astronomers who were more religiously pious and who found it difficult to abandon the medieval perception of a geocentric universe run by supernatural entities.

Galileo paid a price for his defiance, refusing to make his interpretation "just a theory" and not good science. Galileo spent his last years of life in house arrest after being forced to reject his own fact-based solar system. It took the Church almost 300 years to reverse its conviction and clear the record. The Church under Pope John Paul II saw no conflict between the material world and the religious world, as long as the two remained separate. To science is given the material world, including living matter. To religion is given the world of spirit, soul, and God. It is a coexistence that Catholic scientists find acceptable and non-Catholics find tolerable. In Galileo's time, Martin Luther was as much opposed to the solar system as was the Vatican. It took more than a century for the world to ease into a larger universe with a diminished status of earth. We have become a third planet in orbit from the sun, and we are by no means the largest of those solar planets. Our sun is changed into a star, and there are myriads of stars visible to the naked eye in the night sky. With good telescopes, these stars become measured in the billions. When Hubble used his telescopes in the 1920s to look at nebulae, he found most of them to be "island universes" like our own Milky Way. The number of galaxies like our own also became apparent to later astronomers, and they number in the billions.[3] That is very humbling for a worldview based on science compared to the simplistic view of the earth at the center of it all. At the same time, it makes the universe to the pious more mysterious and incomprehensible. It makes us wonder why there are so many galaxies and why the universe is so vast. For some, it makes God's power more impressive. It also makes some people feel diminished and wistfully longing for a return of Dante's universe to return human dignity, but virtually no one seriously believes that Dante's geocentric universe is a plausible one. Science won against an outmoded worldview, but it did not destroy a religion that had once accepted such a view. It would be almost impossible to raise a movement to bring back the geocentric universe once thought integral to religious belief some five or six centuries ago.

This is not the case for Darwin's theory of evolution by natural selection. Almost a century and a half has passed since Darwin presented it. After some initial resistance, the Catholic Church has accepted it, although with ambivalence, and the policy under John Paul's successor, Benedict XVI, some scientists fear, may revert to a compromised evolution that is forced to accept supernatural components (such as Intelligent Design, a concept revived from a moribund Natural Theology that was popular at the start of the 19th century). Most Jews and Protestants accept evolution and do not feel threatened by a worldview that makes a kinship of all life. Human evolution for those who retain a strong religious identification is said by them to be God's way of carrying out creation.[4] The dispute with evolution centers about how a Creator creates. If I told you that my daughter made some pies the other day, none of you would think that she stood over an empty oven, snapped her fingers, and baked pies leaped out. You would assume the verb "make" means going through a process of time-consuming activity. At present, some Evangelical Christians, some Orthodox Jews, and some Orthodox Moslems reject evolution because they reject a materialist interpretation of how life arose or they cannot imagine a God who is a sophisticated scientist using His own laws of nature instead of being a magician using dazzling illusions or miracles. Another century or more may elapse before the worldview of science, as seen through the evolution of life and the universe, is no longer perceived as a threat to religion. At present, evolution makes sense to atheists, agnostics, and those whose religious perceptions do not demand some supernatural explanation for interpreting the universe of life and matter and how it came into being.

The 20th century added two more assaults to the worldview of the past. Einstein introduced ideas of the world of the very large and the world of the very small. He destroyed our belief that matter and energy were separate entities that have no connection to one another. He showed that they were interchangeable. He changed our perception of time and extended Hermann Minkowski's view that time is a fourth dimension. He tied mass, time, and geometry into a common system at the level of the universe. He startled the world by demonstrating that light bends when it passes around the sun or other massive object like a galaxy.[5] Fortunately, most of Einstein's contributions had no apparent connection to religion. The world of the very large and the very small is missing from biblical scripture. There are no atoms or galaxies described in the Bible. In that older universe, the earth is the largest object in the universe, and the stars, planets, moon, and sun are afterthoughts (created after the earth) that may not be material objects. Instead, Einstein's revolutionary ideas made the universe stranger and more remote from common understanding. Supplementing Einstein's views of the universe are the many innovations of 20th century physics, including quantum mechanics, indeterminacy, the four fundamental forces of physics, the expanding (if not accelerating) movement of all galaxies, string theory, dark matter, black holes, and a plethora of atomic particles. They have made the universe stranger still.

The most recent change in our perception of our world comes from molecular biology. When the gene became a molecule that could be described, analyzed, and used chemically, thanks to the Watson-Crick double helix model of DNA, evolutionary biology also went molecular. So did almost every field of the life sciences. It stripped biology of its last hiding place for vitalism, holism, and the supernatural as explanations for living activity.[6] For many scientists, it is a magnificent world of life to contemplate with a past history ready to be read in the genes of our living kin in the world of nature and in the ancient bones, teeth, and spores that still retain vestiges of their DNA. The world of life is perceived by geneticists as an immense repository of molecular events—duplications and shifting of genes, chromosomes, and segments of contiguous genes—that can be analyzed historically and sequentially, as one by one the genomes of our 96 phyla are unraveled and compared.

Science does not destroy the sense of awe; it helps create it and enlarge it, shifting it to always new worldviews that are both strange and beautiful to contemplate because they enlarge our comprehension. These feelings are the traditional experiences provided by literature, religion, and philosophy, and they are more likely to be vitalized by science than to flourish in a stagnant world of knowledge and belief that does not change.

Science Brings About Changes in the Human Condition

The shift from fatalism to causality was made possible by the application of reason to the natural world. That is another way to describe science. It became a major worldview in the 18th century when French and English scientists believed there were no limits to change that science could bring about. That view was forcefully expressed by the French scholar and mathematician, Jean Marie Nicholas Caritat, Marquis de Condorcet.[7] Condorcet used the term "progress" to describe what science could do. This included an eradication of disease, an extension of life expectancy, a subsequent control over reproduction to prevent a population explosion, the end of slavery, universal education with equal opportunity for women to participate in society, and the democratization of the world with an erosion of tyrannies and monarchies. In the long run, Condorcet was right, and his Promethean view of the future was remarkable. He erred in thinking that progress would be more linear than it is. He did not anticipate totalitarian movements, nor did he anticipate world wars. He was, after all, using reason to project its good intentions and accomplishments.

The example I use to illustrate how Condorcet's optimism is justified is the effect of the germ theory in the 1870s on the development of the 20th century in those nations that put it into use. Both Louis Pasteur and Robert Koch identified specific diseases with microbes.[8] Pasteur first recognized this in spoiled wine that turned to vinegar. He noted that wines had small globular bodies which gave way to even smaller non-globular rod-like structures, and he surmised that these microbes shifted the fermentation process of sugar from alcohol to acetic acid. He proved that they could be destroyed by heat, and from this initial insight he developed the procedure for pasteurization of foods, like milk, that spoil. He extended his observations to human diseases. He recognized that weakened microbes (as in dried-out tissue from rabid animals) could be used to make vaccines that immunized persons against disease. That work was inspired by Edward Jenner's earlier use of cowpox to prevent smallpox by vaccination. Koch worked out the techniques to make bacteriology a science. His rigor (encapsulated in Koch's postulates) was essential in demonstrating to skeptics that a particular causative microbe was always associated with an infectious disease and could be transmitted in controlled experiments from laboratory cultures to animal subjects. Both of these schools led to a flourishing germ theory of infectious diseases; over the next half century, almost every known infectious disorder had an identifiable microbe associated with its cause.

The germ theory greatly enhanced the public hygiene movement that was organized efficiently in Germany by Rudolph Virchow and exported to the U.S. and other industrialized countries. Until the germ theory, about half of all infants born died in their first year of life of pneumonia or gastritis, both of which were associated with infectious microbes. The mean life expectancy in the 1880s was about 45 years.[9] The germ theory led to pasteurization of milk, chlorination of water, use of food preservatives, visiting nurse associations for the newborn, and sterilizing treatments of foods in the canning industry. In response, the infants benefiting from these changes brought about by science survived. This led to a population explosion, because in the era before the germ theory, the average number of pregnancies was four to six in the industrial nations, and death in early childhood cut their survival in half, making the world population relatively stable at a replacement level. The sudden elimination of infant mortality by the germ theory now led to a doubling of the population every generation. What Thomas Malthus had feared had now become real. But Malthus relied on disease, war, and starvation (or his preferred choice—sexual abstinence) to limit population growth.[10] Condorcet and his fellow scientists relied on science to solve the problem. That was brought about by the birth control movement (starting with Margaret Sanger in 1913), which used artificial means (mechanical, surgical, and chemical) to limit human reproduction.

Although new diseases can arise, as the pandemic of AIDS has demonstrated, the roster of former diseases that were part of everyone's experience or worry in the 19th century have largely disappeared. We immunize our children against whooping cough, measles, mumps, and polio. We no longer fear outbreaks of yellow fever, cholera, or plague in the developed nations. We have eliminated smallpox and no longer require vaccination to prevent it. Young adults today rarely die of tuberculosis (think of Puccini's opera *La Bohème*) or pneumonia (as in James T. Farrell's Depression-era novel, *Studs Lonigan*). Antibiotics and public health have made it much easier to plan on having two children and expecting those two to live to reproductive maturity. Smaller families (the norm after the birth control movement of the 1920s) have led to more resources and parental time invested in children and a confidence that they will have an educated, fulfilled, relatively healthy life.

What I have demonstrated for the consequences of the germ theory can be applied to science in abundance from chemistry (e.g., plastics, fertilizers, and pharmaceuticals), mathematics (especially computers), and genetics (the green revolution). Physics has supplied radioactive elements for biomedical research, diagnosis, and treatments, as well as atomic energy and a future use of nonnuclear energy sources not dependent on carbon dioxide-producing wastes. None of this, as this book certainly argues, is without some bad outcomes. Those bad outcomes are largely preventable if we include thinking about our values (other than just our good intentions) and the importance of consulting others, especially our critics, in our applications of science. Those bad outcomes are far less frequent than the good that science has accomplished. We cannot dismiss the bad outcomes as the "price of progress," but we can and should act thoughtfully and cautiously as we promote our good intentions.

Notes and References

1. Homer, *The Odyssey*. Translated by Robert Fagles and published in 1996 (Viking Press, New York).
2. Dava Sobel, *Galileo's Daughter: A Historical Memoir of Science, Faith, and Love* (Walker, New York, 1999).
3. Gale E. Christianson, *Edwin Hubble: Mariner of the Nebulae* (Farrar, Straus, Giroux, New York, 1995).
4. Ronald Numbers, *The Creationists: The Evolution of Scientific Creationism* (University of California Press, Berkeley, 1993).
5. Abraham Pais, *Subtle Is the Lord: The Science and the Life of Albert Einstein* (Oxford University Press, New York, 1982).
6. James D. Watson, *DNA: The Secret of Life* (Knopf, New York, 2003).
7. Edward Goodell, *The Noble Philosopher: Condorcet and the Enlightenment* (Prometheus Books, Buffalo, 1994).
8. William McNeil, *Plagues and Peoples* (Anchor Books, Garden City, New York, 1976).
9. Elof A. Carlson, *Human Genetics* (Heath and Company, Lexington, Massachusetts, 1984). See Chapter 1, "The human condition."
10. Thomas R. Malthus, "An essay on the principle of population." In *The Norton Critical Edition of T. R. Malthus,* ed. P. Appleman (W.W. Norton, New York, 1976).

19

How Can Good Intentions Avoid Bad Outcomes?

A S THE 21ST CENTURY BEGINS, science shows no signs of exhaustion. Hundreds of aspects of the universe remain difficult to interpret. In my own field of genetics, we do not know whether the sequenced human genome is really a "blueprint for life." That doubt exists because our genes can make more than one protein by shuffling the exons that are transcribed and translated, and the proteins produced by different genes can then be organized in cooperative complexes leading to epigenetic activity (i.e., factors that do not directly affect DNA) that is not easily reducible to the sequences of genes. Whether there is a yet undiscovered class of genes that coordinate such epigenetic activities of proteins is not known. Clearly, there are switches or regulatory genes (possibly accounting for 3% of the genome) that determine when and where specialized developmental regulators (such as genes bearing Hox boxes) turn on in the development of an embryo. A new field of evolutionary developmental biology ("evo-devo") has mushroomed during the 25 years since E.B. Lewis worked out the evolutionary biology of wing formation in fruit flies. Comparative genomics and "evo-devo" have just begun, and a century from now there will likely be insights into developmental processes, cellular dynamics and structure, evolutionary mechanisms, and signaling among cells that we can only guess at today. What is almost certainly true is that each new finding in our basic understanding of living processes will lead to applications to human health and commercial usage. For the most part, those good intentions will have good outcomes.

The political interference in how science should be supported, taught, or carried out is not new. Both church and state have had their past heavy-handed ways of repressing or tarnishing science. Is there a danger of the elementary school and high school biology courses in the industrialized nations (especially the U.S.) being forced to give equal time and credibility to supernatural creationism in its many forms? This is a real threat, and even without formal approval may change the way science has been taught; many schoolteachers are intimidated by their school principals, their school boards, or their students' parents. It is safer for some teachers to leave out the idea of evolution (or to skim it superficially) and to avoid dwelling on the past history of life on earth. A conscientious effort to present evolution as a well-developed science in elementary or secondary schools could invite ostracism in a religious-dominated community, and the teacher might be forced out of the school system.

Do Contemporary Issues Differ from Those of the Past?

From our analysis of cases through the end of the 20th century, we can classify our concerns:

- The need for regulation in medicine to prevent patient abuse. This includes the shift from paternalism to recognition of the autonomy of adults in making decisions about their treatment. It includes the use of informed consent when conducting experiments with human subjects.

- The need for regulation in the production and marketing of prescription drugs. This includes the objective staffing of federal agencies free from political or commercial pressure so that they can represent the interests of the user, not the manufacturer, of drugs.

- The need for self-evaluation of one's moral standards for those engaged in basic research so that the work is done honestly and with proper controls to prevent self-deception and error. The ostracism and painful publicity of those caught cheating is working well. Most universities and institutions have review boards to investigate charges of fraud. Those found guilty usually resign or are forced to leave.

- The need for regulating the uses of natural resources and the effects that those uses have on the ecology and health of the environment. This includes awareness of polluting agents, inappropriate disposal of wastes, and uninformed use of agents that might do harm to people, wildlife, or the future occupation of the land, water, and air we depend on. It also includes recycling resources and monitoring the environment to detect harmful trends.

- The importance of providing scientists the freedom to explore the material universe without fear that those who disagree will have the authority of the state to alter or deny the findings of those scientists when they try to teach their findings in their courses.

- The importance of keeping the state out of the business of defining what is legitimate science and then using that state-supported science for carrying out its ideological goals. It would be difficult to do this by legislation. Science depends on a vigilant press and the wrath of voters to oust those elected or appointed officials who try to redefine science or to introduce pseudoscience and the supernatural as legitimate science.

- Keeping at bay, as far as it is possible, the influence of religious groups on the federal funding of research, the funding of medical treatments, and the development of new fields of medicine and science because they are seen as immoral by some. This would be difficult to accomplish through legislation. So far, the most effective method to prevent this interference is through the courts or through the electorate. Religions, like business and labor, have political clout, but so do people who vote.

These are issues that will persist well into the 21st century. I do not list the more troublesome (and unquestionably, very harmful) outcomes of the participation by scientists in the development of weapons of mass destruction. I cannot, because in every country this is a matter of patriotism and national security that trumps all other values and can be intimidating to those who challenge the authority of the state to do what it desires, allegedly in the people's best interests. Sometimes it takes centuries for prevailing views to die—like slavery, witchcraft, the perception of women as chattel, the colonization of other people, and the divine right of kings to rule. Waging war may someday join that list.

There are some emerging issues that may take on importance in the 21st century:

- The costs of drugs and medical services and their availability to those who need them. This is a complex issue. As more drugs and procedures become available, as people live into their 90s in the 21st century, and as family size hovers around replacement level in most countries, it becomes difficult to find the funds to keep people healthy who may be retired from work for 30 or more years of their lives.

- As the world population approaches seven billion people (and may rise several billion more during the last decades of the 21st century), the demands on natural resources will lead to multiple problems—global warming, competition for carbon-based fuels, conversion of land into highways and other needs of cars and trucks, loss of land for building homes, potential pollution, and crowding of cities.

- The challenge to supernatural views of the universe and the moral positions of some religions posed by the findings of comparative genomics, the analysis of the primate genomes giving unambiguous evidence of the step-by-step changes accompanying the hominid transformation of the primates into humans, the creation of simple forms of life in test tubes, and the cloning of a few humans for vanity satisfaction.

- Concerns over the patenting of genes and biochemical processes to achieve this, for both natural and modified genes, and the effects this can have on conducting basic research and limiting access to its benefits.

- The potential abuse, by aggressive marketing, of genetic testing for thousands of gene mutations, and the not-well-explored eugenic consequences of wide-scale testing if funded by the state.

Preventing Bad Outcomes

As the tools and products of biological engineering become more abundant and enter medicine, pharmacy, and agriculture, the need for regulation will increase. Regulation works if the agencies have competent scientists, if the agencies are independent of the companies they regulate, and if the regulations are regularly reviewed by objective panels. Despite the fears of the regulated, I expect that the science coming from comparative genomics will vastly reduce the trial-and-error search for effective drugs and vaccines and make their testing easier and cheaper. I also expect that issues like stem cell research will diminish as artificially changed cells from somatic tissue are made into cells capable of taking on specific tissue functions without being fully embryonic in their plasticity. In every new field of science, there is a period of adjustment as techniques become more efficient and circumvent the objections to the first products used.

Certainly, self-regulation is the scientist's ideal. Scientists monitor their own behavior and learn, through years of predoctoral and postdoctoral immersion in research, the value of honesty and careful experimentation to prevent jumping to conclusions prematurely, offering work that is not carefully controlled, and avoiding conflicts of interest. These precepts are essential for good science.

What We Can Learn from History

It is easy to project the present into the past as well as into the future. Very often, what we take for granted in the present was just not done in the same way in the past. Consider patents on drugs. Most people would assume that ever since there have been patents and copyrights for intellectual property rights, there have been patents issued for every drug. This is not so. The idea of patents and copyrights was debated in the first half of the 19th century. Laissez-faire capitalists who believed in competition opposed such efforts, describing them as monopolies in restraint of trade; they argued that governments had no business using their force to take sides. Some argued that the government should, instead, levy a tax on the profits of such companies and distribute this to the inventors. Even when patents and copyrights were authorized, most governments in Europe and the U.S. did not permit the patenting of prescription drugs. They were persuaded that this would lead to monopolistic price gouging which would make needed medicine unavailable to the poor, a result that was considered immoral. Drug companies circumvented that in the early 20th century by patenting, not the product, but the process by which a product was made. It was not until the 1950s that the patent laws in Europe and the U.S. were accommodating to the lobbying efforts of drug companies, and the courts upheld the drug companies. This created a concern among scientists in the 1990s when some companies attempted to patent gene sequences or portions of the genome. Private companies do not have to release such sequences even if a competitor or a research group is willing to pay a premium for them. This procedure can delay research. Fortunately, much of the sequencing of genomes has been done by government-sponsored researchers, and these sequences are in the public domain, so that threat did not materialize.[1]

We know from such historical examinations of a field or product that it is difficult to predict the future and whether the original intentions of a new field or product will turn out as planned. It is also difficult to predict how government regulations, legislation, and industrial policy will react to social pressures, lobbying, and other unforeseen events.

Preventing Bad Outcomes Is Worth Our Effort

Although we cannot write legislation that covers all future contingencies or conceive of a regulatory agency that meets all consumer needs, we can improve the education of scientists so that they become aware that errors have happened in the past. Most scientists would not wish their work to be associated with events leading to death or illness. I believe scientists do have ethical standards and moral feelings which they would exercise if they knew that such bad outcomes had happened in the past and were likely to happen again. This requires exercising those moral values and learning to speak out, to think ahead, and to offer ways to test such doubts. It also requires more oversight by those whose specialties are often lacking among the scientists involved in the design and production of a product or a new process. We cannot expect chemists and physicists to know the environmental consequences of such new products, but we can expect corporations, research institutes, and universities to set up appropriate review boards to consider short- and long-term consequences of the use and disposal of scientific products. One of the remarkable feats of engineering, medicine, and other applied sciences is how well they do learn from experience.

Notes and References

1. I sum up a symposium session held at the International Congress of the History of Science, Beijing, July 24–30, 2005, chaired by Daniel Kovles and Jean-Paul Gaudillière. July 28: "Science and intellectual property in international perspective."

Index

Virchow, Rudolph, 21, 193, 210
Virtue, ethical system based on, 12–13
von Verschuer, Otmar, 25, 28–29, 31
von Wagner-Juaregg, Julius, 141
von Wassermann, August, 141

W

Walsh, Maurice, 201
Wannsee Conference, 23, 31–32, 39
War, as justification for suspending
    ethical and moral behavior
    by scientists, 65–66
War crimes, assessment by international
    tribunals, 33–35
Watson, J.D., 116
Weapons of mass destruction, 188–189,
    215
Weismann, August, 43
Weiss, Ted (Congressman), 91
Weldon, W.F.R., 40
Wells, Herman, 196

Wilmut, Ian, 167
Wilson, E.B., 42
Wilson, Robert, 71
Worldview, changes by science, 205–211
Würtemberg, 32

X

X-linked recessive lethal mutations, 69
X rays
    discovery of, 68
    facts, 67–68
    first uses of, 68–70
    induction of mutations by, 131
    protection from, 77

Y

The Youngest Science (Lewis Thomas), 148

Z

Zumwalt, Elmo, Jr. (Admiral), 91